# Information Retrieval: SciFinder®

# Information Retrieval: SciFinder®

Second Edition

DAMON D. RIDLEY

*School of Chemistry, The University of Sydney*

A John Wiley and Sons, Ltd., Publication

*Registered office*
John Wiley & Sons Ltd, The Atrium, Southern Gate, Chichester, West Sussex, PO19 8SQ, United Kingdom

For details of our global editorial offices, for customer services and for information about how to apply for permission to reuse the copyright material in this book please see our website at www.wiley.com.

*Library of Congress Cataloging-in-Publication Data*

Ridley, D. D. (Damon D.)
    Information retrieval : SciFinder / Damon D. Ridley. — 2nd ed.
        p. cm.
    Includes bibliographical references and index.
    ISBN 978-0-470-71247-4 – ISBN 978-0-470-71245-0    1. Database management. 2. Information storage and retrieval systems—Science. 3. Online bibliographic searching. 4. Science—Databases. 5. Information retrieval. I. Title.
    QA76.9.D3R543 2009
    025.06′5—dc22

                                                                        2009019180

A catalogue record for this book is available from the British Library

ISBN: 9780470712474 (hbk) 9780470712450 (pbk)

Typeset in 10/12pt Times-Roman by Laserwords Private Limited, Chennai, India.
Printed and bound in Singapore by Fabulous Printers Private Ltd

# Contents

# Preface

Since the early 1990s, Chemical Abstracts Service (CAS) has been developing an intuitive 'point and click' interface to scientific information for scientists. Subsequently, SciFinder® was released in 1995 as desktop software for commercial organizations, and the academic version SciFinder Scholar® became available in 1998. Today these are merged into one product, SciFinder, although there are two interfaces: a client version, which requires the software to be downloaded, and a web version. This book focuses on the web version and on how it works, and suggests how scientists may obtain maximum value from the interface.

This book builds on the text 'Information Retrieval: SciFinder and SciFinder Scholar', which was published in 2002. Feedback from that text was positive, but a general comment was 'I don't want to read 240 pages'. I understand. Actually I requested the publishers to market that text as 'about 100 pages, with additional figures and diagrams'. The publishers replied, 'You are the first author ever to ask that a book be marketed with fewer pages that it actually has!'

The biggest changes in the information world since 2002 have been the dramatic increase in availability of full text documents in electronic format and the increased search capabilities on web interfaces. Other changes relevant to scientists have been the introduction of Reaxys and Scopus, and enhancements to Web of Knowledge. Having studied these options, I still am of the opinion that SciFinder offers the best package – provided that users understand how to use it effectively. I hope this books helps.

While it is understandable that scientists really want an interface that requires minimal learning, the reality is that there are numerous issues relating to retrieval of scientific information. Issues start with the content ('What really is the data behind this interface?'), then go through the difference between author text and indexing ('Actually what is indexing and how do I learn it'), to the opportunities the search engine provides both at the search and then at the post-processing level. It's complex, but SciFinder offers simple solutions.

SciFinder is very different from all other information retrieval tools. The most user-friendly alternative search tools, including web search engines, still require some knowledge of truncation, of proximity searching, of synonyms, and so forth. They interpret the question literally and rarely do they offer any guidance on how to proceed. Depending on what the scientist really wanted, they may or may not produce comprehensive and/or precise answers.

SciFinder also works differently in that it *guides the searcher*. Indeed, when a question is initially asked, SciFinder does not give a straight answer! Instead, it guides the searcher by producing a set of options. SciFinder tells the user to 'go down this option and you will find a specified number of records, whereas this other option provides a different number of records'. In other words, the user chooses a path based on the actual number of hits, but the choice is not irrevocable and the user may always return to narrower or broader answer sets.

Once a path has been chosen, SciFinder has *creative analyze options*. For example, for initial bibliographic answers SciFinder provides histograms of the different index terms, or document types, or author names, or publication years. Armed with a lot of information up-front the scientist chooses refinement paths accordingly. In chemical structure searching, SciFinder automatically interprets the query to allow for different structure conventions and representations. If answers are too numerous, then SciFinder again guides the searcher through analysis tools for substances.

This text is not designed to explain the mechanics of searching or of data processing. These are detailed in the numerous help messages available through the help icons in SciFinder.

Instead this text explains what 'goes on behind the scenes' and gives examples of search strategies. The text starts with an outline of the basic content of the databases and the way SciFinder searches these databases. It explains why certain answers are retrieved and how features of SciFinder may be used to narrow or broaden searches in a predictable way. It explores different options to the solutions of problems, and above all it encourages scientists to be creative and to think carefully about how to approach problems. SciFinder is a *research tool* and not just a search tool!

The searches in this text were conducted in the early part of 2009, and the SciFinder functions and types of answers obtained were current at that time. Of course things change, and significant changes will be posted as appropriate at www.wiley.com/go/ridley_scifinder. Additionally, this url links to a number of exercises and it includes many of the SciFinder screens that appear in this book.

The SciFinder concept is highly innovative ... indeed brilliant! The implementation of the concept took years to achieve, and resulted in combined efforts of CAS staff and scientists worldwide. Particular thanks are due to CAS management and staff at CAS for making it possible. Over the last 25 years it has been both a privilege and enjoyable for me to be associated with CAS as a consultant, educator, advisor, and most importantly as a scientist.

My work with CAS has taken me to over 30 countries, and this book is dedicated to all the people I have met in my travels ...

... and to those who have helped me 'at home', particularly Eva; Lloyd, Andrew, Nicholas, and Natasha; Fred and Julia; William, Matthew, Amy, Josie, Claire, Daniel, Lucy, Emma, and Oliver.

Damon Ridley
April 2009

# 1

# SciFinder®: Setting the Scene

## 1.1 'I Just Want to Do a Quick and Simple Search on ...'

... is sometimes heard in scientific laboratories. It can be achieved, provided the scientist has the background knowledge, but the catch is that getting this background knowledge may not be 'quick and simple'.

We will discuss background knowledge later, but to get started consider a 'quick and simple' search on inhibitors of $\alpha$-carbonic anhydrase from *Helicobacter pylori* (a topic relating to the Nobel Prize in Medicine in 2005). Several questions immediately come to mind including:

- In what database(s) should we search? Should we be searching SciFinder®?
- What search terms should we use? Will they give comprehensive and/or precise answers? Are we likely to miss important answers, and why?
- Assuming we get a large number of answers, how will we narrow them? Or if we get few answers, how may we make our search more comprehensive?

We will briefly work through these questions.

### 1.1.1 Databases

The term *database* may mean some collection of electronic documents or records including:

- Documents available on the Internet;
- Full text collections (e.g. journals or patents);
- Abstract and indexed (A&I) databases.

We access this information through a variety of *search interfaces*; some well-known interfaces used by scientists include Epoque, Google™, Reaxys®, ScienceDirect, SciFinder®, Scopus®, Web of Science®, Wikipedia®, and Wiley-Blackwell.

*Information Retrieval: SciFinder®, Second Edition* Damon D. Ridley
© 2009 John Wiley & Sons, Ltd

On the other hand, the names of *A&I databases* are less well known although some scientists may be familiar with databases such as BEILSTEIN, BIOSIS®, CAPLUS^{SM}, COMPENDEX, EMBASE, MEDLINE®, CAS REGISTRY^{SM}, and SCISEARCH®. When choosing databases to search it is important to understand their content and the opportunities that their search engines provide. For example, full text databases require different search techniques from those in A&I databases, and while some A&I databases cover similar collections of journals nevertheless the content of the records in the databases may be very different – and searches need to be adjusted accordingly.

To understand these issues requires considerable study, but at this stage suffice it to say that the SciFinder interface and its major databases (CAPLUS, MEDLINE, and REGISTRY) offer many advantages over all other interfaces – provided the searcher understands and uses these advantages. Understanding how to make use of these advantages is the point of this book.

---

*Comment*

Different search interfaces and databases require different search strategies, and offer different options for narrowing or broadening initial answers. We need to understand what we are searching and how to search it!

We search database content, and we need to know whether it is a primary source (journal or patent) or a secondary source (an indexed database). We need to know what years are covered and what are not covered. If the database contains indexing then we need to know a little about index entries.

How we search depends on the content, but also on how the search engine works. Does it allow for truncation, does it interpret the question exactly or does it use algorithms to enhance the search, and what opportunities are there to post-process answers?

---

### 1.1.2   Search Terms

The search question above has three concepts (*inhibitors*, *α-carbonic anhydrase*, and *Helicobacter pylori*) and each needs to be considered separately. While there are many considerations (e.g. singulars/plurals, different spellings, single terms that mean different things, synonyms) relating to choice of search terms for concepts, often the major task is to find synonyms.

Documents reporting *inhibitors* are likely to contain text using terms like inhibit, inhibition, inhibiting, inhibitor(s), all of which would be covered through use of truncation (e.g. inhibit? or inhibit* – it depends on the truncation symbol the search engine recognizes). Additional synonyms in this case are few, but the major problem is that there are many classes of, and many different names for, inhibitors. Finding search terms for these would be very difficult.

The Greek letter 'alpha' in *α-carbonic anhydrase* presents one issue, but the main issue is to cover the synonyms, which may include carbonate dehydratase, carbonate anhydrase, or even EC 4.2.1.1. We simply do not know which term different authors may have used, so to be reasonably comprehensive we need to search them all.

Those in the field know the main terms for the *bacterium* are *Helicobacter pylori* or *H. pylori*, so this concept may easily be searched. However, there are over 50 different *Helicobacter* species known and the question is whether or not some of these may be of interest as well.

The recommended search strategy is to search a few concepts initially, particularly those for which reliable search terms may be chosen. In this case it would therefore be advisable to start the search with terms for the enzyme and the bacterium. Depending on the outcome, various strategies may then be used to find documents on the inhibitors.

---

*Search Tip*

Make sure synonyms for terms are included – no matter what search system is used. However, a single term may have multiple meanings (e.g. cell in bacterial cell or cell phone), so increasing the number of synonyms may also increase the number of false hits through retrieval of entries where different word meanings apply.

False hits of this type may be removed by addition of more concepts (e.g. adding the search concept 'bacterial' to the term 'cell'), but the more concepts added the more synonyms that need to be considered. Therefore, start with a few concepts for which few synonyms apply, then look through answers, and narrow or broaden them in systematic ways.

---

### 1.1.3   Narrowing Answers ⇒ Precision; Broadening Answers ⇒ Comprehension

Having obtained our initial answers we may find:

- Too many answers and perhaps some that are irrelevant (e.g. because the terms searched have alternate meanings);
- Surprisingly few answers, in which case we may be concerned that we are missing relevant information;
- A manageable number of answers, but numbers of answers alone are far less important than answer precision or comprehension.

No matter what initially turns up we invariably need to take further steps. In some cases (e.g. in a Google search) we may simply ignore answers beyond the first couple of screens, while in other cases we may go through answers manually and mark those of interest. We all have experienced this situation and no doubt we have developed our own strategies to proceed.

The important things are that we make rational (scientific!) decisions that we understand what we have done and that we know the types of answers we have included and excluded. Sometimes the key driver is the number of answers: too few and we risk missing data; too many and we may spend too much time working through a large volume of material.

## 1.2    The SciFinder Way

We will look at options to get initial answers through SciFinder and then at options to narrow them in systematic ways. Without going into too many details we will simply make ourselves familiar with the SciFinder process. It soon becomes apparent that SciFinder is a unique tool, with content and functionality well in advance of other search interfaces.

### 1.2.1    Getting the Initial Answers

After signing in to SciFinder the default **Explore References: Research Topic** screen appears (Figure 1.1). While there are many other explore possibilities, we will focus on the **Research Topic** query box and on the examples under it.

These examples suggest natural language statements and in particular we note the use of the prepositions 'on' and 'of'. The inclusion of prepositions actually has a significant

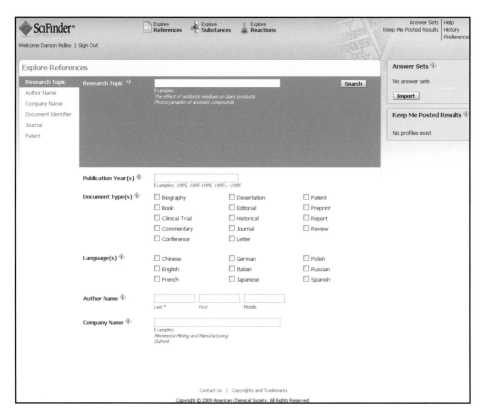

**Figure 1.1**    *SciFinder's* **Explore References: Research Topic** *screen. Other* **Explore References** *(bibliographic) options such as* **Author Name** *and* **Document Identifier** *are on the left and entries to* **Explore Substances** *and* **Explore Reactions** *are at the top. Additional bibliographic search options are at the bottom. SciFinder*® *screens are reproduced with permission of Chemical Abstracts Service (CAS), a division of the American Chemical Society*

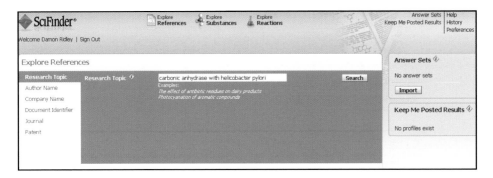

**Figure 1.2**  *Data entry in **Explore References: Research Topic** query box. Note the use of the preposition 'with'; avoid use of Boolean 'AND'. SciFinder® screens are reproduced with permission of Chemical Abstracts Service (CAS), a division of the American Chemical Society*

impact on the types of results we obtain, but we will talk about that later. Right now, we will follow this lead and enter 'carbonic anhydrase with helicobacter pylori' (Figure 1.2), and then click **Search**.

We obtain the screen (Figure 1.3) that is very different from the next screen we would obtain through virtually all other Internet search interfaces, notably:

- *It's not an answer set at all*;
- SciFinder has interpreted our query, has looked through its databases, and has come up with some alternatives.

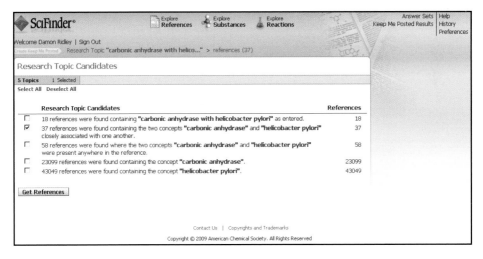

**Figure 1.3**  *Research Topic Candidates screen.  SciFinder has considered the query (Figure 1.2), has applied a number of search algorithms, and has presented some alternatives for consideration.  The list of options becomes more general from top to bottom. SciFinder® screens are reproduced with permission of Chemical Abstracts Service (CAS), a division of the American Chemical Society*

SciFinder is simply guiding us. SciFinder:

- Indicates that there are 18 records for the phrase 'carbonic anhydrase with helicobacter pylori';
- Has found two 'concepts';
- Indicates there are 37 records where the two concepts are 'closely associated' (which broadly means in the same 'sentence');
- Indicates there are 58 records where the two concepts are present in a record (Boolean 'AND' has been interpreted);
- Indicates the number of records for the individual concepts.

SciFinder has gone from a precise interpretation to a quite general interpretation of the query, and our choice depends on our requirements. How SciFinder identifies a concept, and the meaning of a concept, is explained in Chapter 3, but a good compromise now may be to click first the box next to the 37 references, then click **Get References**. We obtain the information shown in Figure 1.4.

This screen has many similarities to displays of answers from other internet search engines. We see titles and abstracts, there are boxes on the left for checking and hence

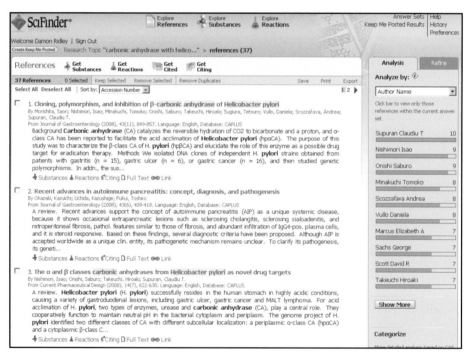

**Figure 1.4**    *Initial SciFinder answer screen (only the first three answers are shown here). Note the links to **Get Substances, Get Reactions, Get Cited**, and **Get Citing** towards the top, and note the **Analysis** and **Refine** options on the right. (**Analyze by Author Name** is the default display but an additional 11 options are available.) These are important functions to expand or narrow answers in systematic ways. SciFinder® screens are reproduced with permission of Chemical Abstracts Service (CAS), a division of the American Chemical Society*

for selecting for other actions (e.g. printing), and the title is hyperlinked to the full database record. We may link to full text, we may retrieve citation information, and we may sort answers. Most of us are familiar with options like these.

However, the SciFinder display also has important additional options: **Get Substances, Get Reactions, Analysis, Refine**, and **Categorize**. *These are the tools that put SciFinder post-processing and explore options at a completely different level from nearly all other internet search options*. If we don't know about, and aren't using, these tools then we are not using SciFinder effectively!

### 1.2.2  Beyond the Initial Display: CAPLUS and REGISTRY

When we click the title for an answer, we see the full record. An example is shown in Figure 1.5.

Bibliographic information for the original article appears on the top right and under Accession Number appears: CAPLUS. This is the name of the Chemical Abstracts Service (CAS) bibliographic database, which is the *world's largest A&I database* for chemistry and related information. Title and abstract information is presented, and while this contains a large amount of information, it is important to recognize that it is usually written by authors – who may not use systematic terms. Indeed the biggest issue with searching the so-called free text is its *variability*. One author may use very different terms from another in the same field, and accordingly searchers may need to use a number of different terms and synonyms to retrieve all the relevant articles.

Many database producers add value by the addition of indexing. The advantage is that indexing is systematic, so a single Index Term may be used to cover many author variations for the topic. However, indexing is a complex area in itself and most scientists know relatively little about it. As we will soon see, SciFinder overcomes this limitation in a number of ways.

Some general comments about the indexing in this record are:

- At the broadest level, this record is in Section 10-2 (Microbial, Algal, and Fungal Biochemistry) and more information on this level of indexing may be found through clicking the 'i' in the blue diamond adjacent to the entry in the record. CAS sections may be searched separately in SciFinder (see Chapter 3, Section 3.5.5);
- **Concepts** start with Index Headings, and this record has headings Stomach, Organelle, Cytoplasm, Mouse, Virulence (microbial), Mus musculus, pH, and Helicobacter pylori. The presence of the last one in the record is significant; the search term we used matches with a precise Index Heading, so we have a high level of confidence that this search term would retrieve records for original articles where an important part of the new science referred to the organism;
- Index Headings are linked, and when the link is clicked SciFinder retrieves all records that contain the Index Heading. This allows us to retrieve records on the same specific topic. If we clicked Helicobacter pylori we would retrieve all records in CAPLUS that were indexed with this heading;
- Sometimes Index Headings have Subheadings, which enable more precise retrieval. For example, the record in Figure 1.5 contains the Index Heading: Stomach, and the Subheading: mucosa. SciFinder provides us with several ways to find Index Headings and Subheadings easily, but we will learn more about this later;

***Figure 1.5*** *Bibliographic answer display (CAPLUS) in SciFinder. Bibliographic information appears on the right. Title and abstract appear above the Indexing which is additionally divided into **Concepts** and **Substances**. Citations appear for documents from mid-1990s onwards. There are 69 cited documents in this record (only the first four are shown). SciFinder® screens are reproduced with permission of Chemical Abstracts Service (CAS), a division of the American Chemical Society*

- After most Index Headings appears some text, sometimes related to terms in the title of the document. Such text is referred to as a 'text-modifying phrase', which we may consider as author-related text and more closely associates that part of the original document with the Index Heading. Recall that we chose from Figure 1.3 those records in which carbonic anhydrase was 'closely associated' with Helicobacter pylori; *inter alia* this restricts records to those in which both concepts appear in the title, in a single sentence in the abstract, or in the Index Heading and its associated text-modifying phrase;

- **Substances** for which important discoveries are reported in the original article are also indexed as a series of numbers (CAS Registry Numbers®) and in many cases a name for the substance appears;
- In the box after the CAS Registry Numbers appears an entry 'Biological study, unclassified'. This entry, called a CAS Role, indicates the type of study associated with the substance and is a precise level of indexing;
- Note the CAS Registry Number 9001-03-0 for carbonic anhydrase has a link and the record for this link is shown in Figure 1.6.

The points listed above apply to the indexing in this particular record in CAPLUS. There are other indexing issues, which we will consider later.

---

*Search Tip*

We should always consider indexing in performing our searches. There are many ways to find relevant indexing 'quick and simple' in SciFinder, and we will see some options in this chapter.

  Why should we consider indexing? There are many reasons, but primarily because indexing is systematic and indexing greatly facilitates searches at all levels from precision to comprehension.

---

CAS Registry Numbers are systematic *index entries for substances*. They appear in all CAS databases and are also used widely in the sciences. The authoritative list of CAS Registry Numbers is the CAS Registry database (REGISTRY), and the display in SciFinder (Figure 1.6) is from this database.

  Each REGISTRY record contains a single substance, various names used for the substance, its structure if available, the CAS Roles listed for the substance in the CAPLUS bibliographic record, and other reported data including spectroscopic data, spectra, other experimental physical and chemical properties, and calculated properties.

  In Section 1.1.2 we mentioned the need to consider various synonyms for substances, and names for the enzyme are shown in Figure 1.6. A search on the CAS Registry Number 9001-03-0 in CAPLUS retrieves all records in this bibliographic database in which 9001-03-0 is entered; since the index policy is to enter CAS Registry Numbers in CAPLUS records where original documents report new science on the substance, irrespective of the name used in the original documents, this search offers a comprehensive and precise route to records.

---

*SciFinder Tip*

Comprehensive and precise searches for substances should nearly always use CAS Registry Numbers! However, sometimes several CAS Registry Numbers, e.g. for a substance and its salts (Chapter 4, Section 4.4.3), may be needed.

---

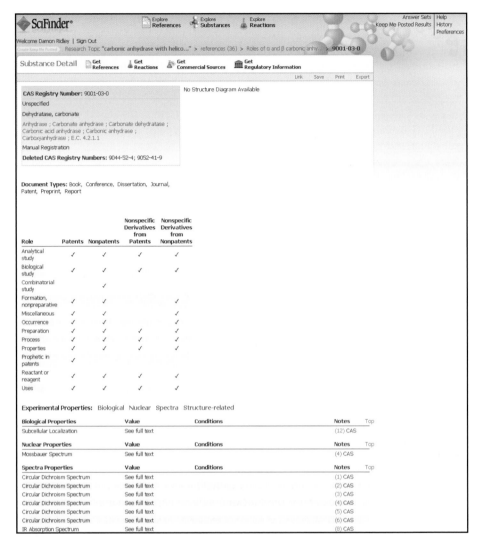

***Figure 1.6*** *Substance answer display (REGISTRY) in SciFinder (only part of the substance record is shown). CAS Registry Number and substance names appear first, then molecular formula and structure if available. Next follows an indication of the types of information that appear for the substance in the bibliographic database. Extensive theoretical and experimental property information follows. SciFinder*® *screens are reproduced with permission of Chemical Abstracts Service (CAS), a division of the American Chemical Society*

### 1.2.3   Beyond the Original Display: MEDLINE

The record in MEDLINE for the same original article is shown in Figure 1.7. There are some minor differences from the record in CAPLUS (Figure 1.5) in the title, abstract, and bibliographic information, and while these can have implications in advanced searches we need not concern ourselves with them now.

*Figure 1.7* *Bibliographic answer display (MEDLINE) in SciFinder. The layout of the information is similar to that given in Figure 1.5. SciFinder® screens are reproduced with permission of Chemical Abstracts Service (CAS), a division of the American Chemical Society*

The important differences are in the indexing. These occur because different organizations (Chemical Abstracts Service and National Library of Medicine (NLM)) produce these databases and the database producers may view documents in different ways.

Some general comments about the indexing in this MEDLINE record are:

- Index Headings may be inverted (e.g. Mutagenesis, Insertional) so immediately we are alerted to the fact that we have to be careful searching for specific phrases (e.g. a search in MEDLINE on the exact phrase 'insertional mutagenesis' will not retrieve records with the Index Heading);
- Some Index Headings are followed by two letter codes (e.g. ME, GE) with full terms (e.g. metabolism, genetics). These 'allowable qualifiers' enable us to search for more specific information relating to the Index Heading;
- Different CAS Registry Numbers from those in Figure 1.5 are indexed (remember, these records are for the same original document);
- Chemical names appear and some are at general class levels (e.g. Acids) while others are quite specific (e.g. EC 4.2.1.1).

SciFinder has tools to help us understand indexing in CAPLUS and in MEDLINE, and we will consider these later. It is relevant now simply to note that there are differences in the indexing systems.

### 1.2.4  Post-processing: Analyze/Analysis

The current answer set has 37 records, and one option is to look through the answers in turn. Even with such a small number of answers this may be time-consuming.

Further, these 37 records were chosen from initial candidates (Figure 1.3) simply in order to understand something about the types of records on this topic in the database. Almost certainly initial answers are neither comprehensive nor precise, and ways to broaden or narrow answers need to be considered. Therefore, we need to do some post-processing, and the most used post-processing tools (tools to narrow or broaden initial answer sets) are visible through the display on the right in Figure 1.4. They are **Analyze, Refine**, and **Categorize**.

Through **Analyze** we ask the search engine to do all the hard work for us, specifically to look through all the answers and give us indications of the numbers of records in various areas. The 12 **Analyze** options available for bibliographic records in SciFinder may be viewed by clicking on the down arrow on the right of Author Name (Figure 1.8).

When we move down any of these and release the mouse the computer looks through all the answers and gives a histogram of terms. For example, if we go down to Index Term (we already know that understanding indexing is important) we see the start of a histogram of Index Headings (Figure 1.9). Click **Show More** and all Index Headings appear (Figure 1.10). We now need to work out what all this means!

We know that this is a histogram of Index Headings, but they seem to be duplicated (e.g. Human, Humans); some are preceded by an asterisk (*) and some have two letters after a colon (:). First, remember that the initial answers are from two different bibliographic databases (CAPLUS and MEDLINE) and we have already seen (Sections 1.2.2 and 1.2.3) that the different databases have different indexing. It just happens that,

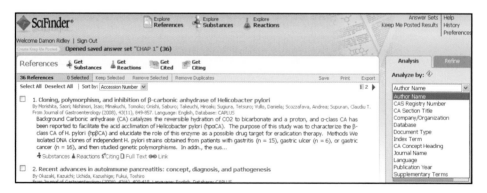

**Figure 1.8**  *Analysis options for bibliographic records. Through these options SciFinder searches all answers and presents a histogram of terms. SciFinder® screens are reproduced with permission of Chemical Abstracts Service (CAS), a division of the American Chemical Society*

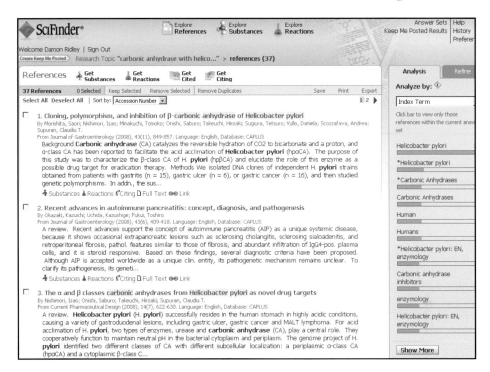

**Figure 1.9** *Initial screen for **Analyze: Index Term**. The most frequently occurring terms appear and clicking **Show More** (bottom right) provides the full list. SciFinder® screens are reproduced with permission of Chemical Abstracts Service (CAS), a division of the American Chemical Society*

for example, the Index Heading in CAPLUS is Human and in MEDLINE is Humans. MEDLINE indexing also indicates the most significant headings with an asterisk, and the text after a colon relates to MEDLINE allowable qualifiers (secondary index levels). So we can quickly interpret the different headings (Figure 1.10) and of course we can easily go through the entire list if needed.

We immediately learn some very important items, just from the display in Figure 1.10:

- *Carbonic Anhydrases and Carbonic Anhydrases appear. The characters such as * and :ME are give-away indications that these are MEDLINE headings;
- Helicobacter pylori and *Helicobacter pylori suggest CAPLUS and MEDLINE headings respectively;
- There is an Index Heading Carbonic Anhydrase Inhibitors.

The first two points indicate that our quick and simple search 'carbonic anhydrase with Helicobacter pylori' would have covered Index Headings (provided that singulars and plurals are searched automatically), while the third point suggests a great way to cover systematically the 'inhibitor' requirement in the original question. The Search Tip in Section 1.2.2 promised that indexing is easy to understand in SciFinder and surely here is the proof.

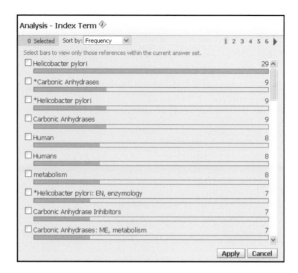

**Figure 1.10**   *Screen providing access to the full list of Index Terms. The screen appears after **Show More** is chosen in Figure 1.9. By default, terms are sorted by Frequency. The screens show the ranking of Index Headings in CAPLUS/MEDLINE. SciFinder® screens are reproduced with permission of Chemical Abstracts Service (CAS), a division of the American Chemical Society*

### 1.2.5   Post-processing: Categorize

While **Analyze: Index Term** is extremely powerful, when applied to large answer sets (where each record may have several different index headings) the analysis list may be very lengthy. Since we need help particularly with large answer sets, we may get into another time-consuming difficulty in the use of this tool in these circumstances. The solution is **Categorize**!

**Categorize** first looks through all the Index Headings and sorts them into predefined categories. Next it sorts headings into subcategories, and finally the specific indexing in each of these subcategories may be displayed. To initiate this process, click **Categorize** on the right-hand side of the display in Figure 1.4 and within a few seconds SciFinder displays the screen (Figure 1.11).

Category Heading **All** is highlighted by default (but we may click any of the other narrower categories if required and follow similar processes to those we will now consider). We note there are two options under the column headed Category:

- Substances, which focuses on the substances in the answer set;
- Topics, which focuses on the Index Headings in the answer set.

When we click **Topics** the Index Headings are displayed (Figure 1.12), and terms we choose are placed in the last column. Finally, we click **Refine** and retrieve an answer set, which has been narrowed through use of the systematic terms chosen.

The value of this process is immediately apparent. We start with an initial search on two of the concepts and then 'quick and simple' we may drill down in very specific ways.

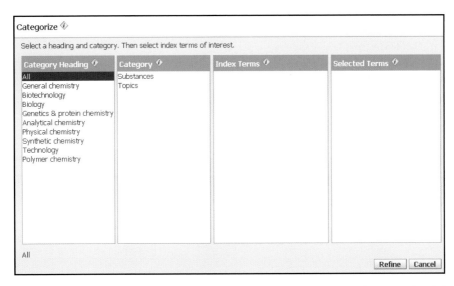

**Figure 1.11**  *Initial screen for **Categorize**. Category Headings are in Column 1. The default display is **All** and Categories within this default are shown in Column 2. (Other Category Headings have different Categories.) SciFinder® screens are reproduced with permission of Chemical Abstracts Service (CAS), a division of the American Chemical Society*

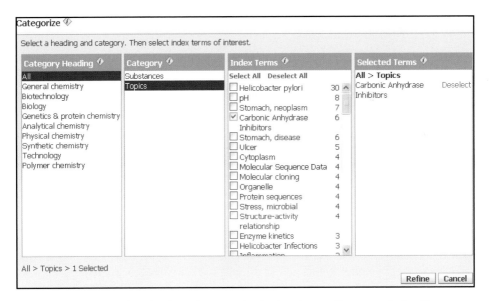

**Figure 1.12**  ***Categorize** display where **Topics** is the chosen Category. Index Terms now appear in Column 3 and terms then selected are shown in Column 4. SciFinder® screens are reproduced with permission of Chemical Abstracts Service (CAS), a division of the American Chemical Society*

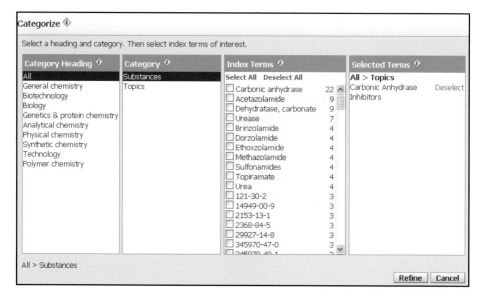

**Figure 1.13** **Categorize** *display where* **Substances** *is the chosen Category. Index Terms now are substances. Some will be inhibitors, while others will be substances reported in the original documents (e.g. urea and 121–30–2 (chloraminophenamide)). SciFinder® screens are reproduced with permission of Chemical Abstracts Service (CAS), a division of the American Chemical Society*

While already we have an indication of how to approach the inhibitors in the original question, we have another option through Category Heading **All**, Category **Substances**, and **Index Terms** (Figure 1.13). A number of substances appear in Column 3, and some of these may be actual inhibitors of the enzyme.

### 1.2.6   Post-processing: Refine

Another commonly used post-processing tool is **Refine**, and the options are shown on the right-hand side in Figure 1.14. We make a direct entry; for example at this stage for **Research Topic** we may enter 'inhibitor' and we would narrow answers to those with this concept anywhere in the record.

The main difference between **Analyze** and **Refine** is that through the former process SciFinder suggests options for us based on what it finds in the databases. This helps to uncover alternatives that perhaps we had not considered. On the other hand, we need to enter terms directly under **Refine** and as we will see later, this process has special uses.

### 1.2.7   Broadening Answers: Get Substances, Get Reactions, Get Citing, and Get Cited

There are 12 options under **Analysis** and seven under **Refine**. There are several additional options under **Categorize**. All of these offer options to *narrow* answers. On the other hand **Get Substances, Get Reactions, Get Citing**, and **Get Cited** (shown towards the top of Figure 1.14) offer options that *broaden* answers:

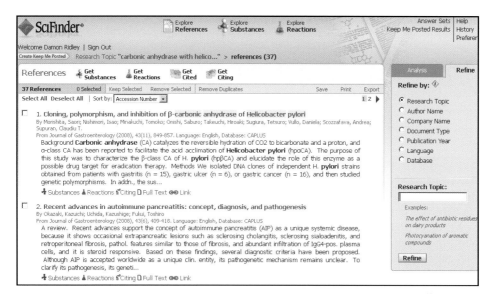

**Figure 1.14  Refine** *options for bibliographic records.  Default is* **Research Topic** *and the Research Topic query box appears. Different query boxes appear when other* **Refine** *options are chosen. SciFinder® screens are reproduced with permission of Chemical Abstracts Service (CAS), a division of the American Chemical Society*

- **Get Substances** displays all the substances indexed in the answer set. If we think about it, if we can find all the substances in a search 'carbonic anhydrase with Helicobacter pylori', then surely some of these will be the inhibitors we seek;
- **Get Reactions** displays all reactions indexed from the original documents.  If we think about it, perhaps not so important for this topic, this option would offer many opportunities for searches where chemical reaction information is of interest;
- **Get Cited** finds records in CAPLUS and in MEDLINE that are cited in the original document.  If we think about it, if documents were cited by the authors they must be related to the science discussed, and finding these cited records may enable us to broaden our search retrospectively;
- **Get Citing** finds records in CAPLUS that reference the documents in the initial answer set.  If we think about it, we now have an option to look at more recent research on the topic.

In the last two cases, bibliographic records are obtained and these may be post-processed in ways similar to those just described for the initial answer set. This helps narrow cited and citing documents to the specific topics of interest.

### 1.2.8  Databases in SciFinder

Currently there are six databases in SciFinder (Chapter 2). CAPLUS and MEDLINE are bibliographic databases in which a single record relates to a single original article. As of January 2009 there are more than 30 million records in CAPLUS and more than 18 million records in MEDLINE. However, the number of records needs to be

**Table 1.1**  *Summary of numbers of records in CAPLUS and in MEDLINE (January 2009)*

|  | CAPLUS | MEDLINE |
|---|---|---|
| Start date | 1907[a] | 1950 |
| Number of records | >30 million | >18 million |
| Number of records with abstracts | >28 million | >10 million |
| Number of records with full indexing | >27 million | >18 million |
| Number of records for patent families | >6.5 million | n/a |

[a]Plus more than 134,000 pre-1907 records.

interpreted with caution since the content of records is also a factor. For example, while most records in CAPLUS have abstract text, the number of records with abstracts in MEDLINE is lower (Table 1.1). In turn this can affect the number of searchable terms in the different databases.

A third database is REGISTRY, which is a database where each record relates to a single substance. REGISTRY is the master collection of disclosed chemical substance information and contains all types of substances, sequences, and natural and synthetic materials, including:

- Elements and subatomic particles and their isotopes;
- Organic and inorganic substances and their salts;
- Alloys, ceramics, and polymers;
- Proteins and nucleic acids including their sequences, primers, and derivatives.

REGISTRY is easily the largest substance database in the world, and as of January 2009 contains more than 42 million organic and inorganic substances and more than 60 million sequences.

Two other databases, CHEMLIST and CHEMCATS, are substance databases that contain regulated and commercially available substances respectively. The sixth database is CASREACT, which is a database that focuses on chemical reaction information.

All databases are linked, which allows information from one database to be transferred to another. For example, we may perform a structure search in REGISTRY, obtain a number of substances, and then easily get references to these substances in CAPLUS and in MEDLINE. Once we have these references we may use the SciFinder tools **Analyze, Refine**, and **Categorize** outlined above.

Alternatively, we may first obtain bibliographic records, obtain the indexed substances and then find all the substances in REGISTRY. SciFinder also has powerful **Analyze** and **Refine** tools to allow post-processing of answers (which of course are substances) in REGISTRY. We may interchange information in and out of CASREACT, and when in CASREACT many **Analyze** and **Refine** tools are also available.

These processes use elegant tools that we need to understand. However, the key to maximizing use of SciFinder is to understand when to use these tools. For example, if we are interested in chemical reactions then when is it better to start in CASREACT, or in CAPLUS, or even in REGISTRY?

Our options expand as we learn more about SciFinder, its content, processes, and tools, and when to use them most effectively.

## 1.3 Looking Ahead

Scientists have vast information resources at their desktops, including the Internet, Google, Scopus, Full Text, Web of Science, many A&I databases – and SciFinder.

However, there is something different about SciFinder. Not only does it contain some of the world's most important databases, but also it contains functionality to allow innovative solutions to information needs. **Analyze, Refine**, and **Categorize** open new horizons as search tools.

In this chapter we have seen many of these functions. Subsequent chapters describe:

- The content of the SciFinder databases in more detail (Chapter 2);
- How to perform text-based searches and to narrow answers in systematic ways (Chapter 3);
- Ways to find information on substances (Chapter 4);
- Issues with substructure searches and how to use substructure searches effectively (Chapter 5);
- Additional search and display options based on bibliographic data such as authors and companies, and citation information (Chapter 6);
- Additional search strategies such as when to use **Explore References: Research Topic** or **Explore Substances**, search issues in CAPLUS and in MEDLINE, and how to search for biological substances and polymers (Chapter 6);
- How to search for chemical reaction information (Chapter 7).

# 2

# Databases in SciFinder

The most important question is 'Does SciFinder meet my information needs?' To answer this, the user first needs to know a little about the content of the databases and then needs to understand how to use SciFinder effectively. This chapter gives a basic description of the databases and subsequent chapters focus on how to use SciFinder effectively.

SciFinder contains five CAS databases (CAPLUS, REGISTRY, CASREACT, CHEM-CATS, and CHEMLIST) and the NLM bibliographic database MEDLINE. There is considerable information available about these databases elsewhere (see Appendix 1 for links).

## 2.1 CAS Bibliographic Database (CAPLUS)

### 2.1.1 Content and Coverage

Up-to-date information on the content and coverage of CAPLUS is available in Appendix 1. The title of this webpage: 'Worldwide coverage of many scientific disciplines all in one source' is an understatement; the reality is that by most criteria CAPLUS is the world's largest and most comprehensive A&I database in the sciences.

Currently the database contains records from more than 30 million original publications from a variety of primary sources dating back to 1907, with some references back to the early 1800s. While the major source is journal articles (70%), there is extensive coverage of patents (20%), reviews (7%), and conferences (4%); among the remaining document types are dissertations, reports, and books.

Some general points are noteworthy with respect to technical content, including:

- Records have technical information in the titles, abstracts, and indexing;
- Over 93% of records have abstract text;
- Over 90% of records have full indexing;
- Titles and abstracts may be enhanced by CAS, which applies in particular to records for patents.

*Information Retrieval: SciFinder®, Second Edition* Damon D. Ridley
© 2009 John Wiley & Sons, Ltd

The total searchable data in all these fields confirms the unparalleled content of the database.

## 2.1.2    Indexing in CAPLUS

One of the benefits of searching databases is that users may take advantage of the indexing added by the database producer. The advantages of indexing are many, including:

- Systematic terminology so that the different terms used by authors to cover a particular topic may effectively be searched under a consistent term for that topic. In this way use of a few index terms will produce more comprehensive answer sets;
- Precise entries into topics, since the index term is entered only if it relates to a focus of the original research;
- Additional entries to the record. For example, not only does the systematic indexing of chemical substances enable precision searches to be accomplished, but also important substances mentioned in the article will be entered in the indexing even though no mention of them may be made in the title or abstract written by the author.

CAPLUS indexing appears in three main categories. First, the database is divided into 80 sections (see Appendix 1 for links) and each record is given the section and subsection number that is considered to be most relevant to the overall content of the article. For the record, in Figure 1.5 the Section Code is 10-2 (Microbial, Algal, and Fungal Biochemistry).

---

*SciFinder Tip*

Section codes are useful for the refinement of initial answer sets into *broad research areas*. This is done by **Analyze: CA Section Title** and is discussed in Chapter 3, Section 3.5.5.

---

The remaining two indexing categories relate to subjects and to substances, and are shown in SciFinder displays under the headings **Concepts** and **Substances** (Figure 2.1). The main issues to understand when using indexing are:

- Indexing may change with time, so an index term may apply only for a certain time period;
- Indexing is different in different databases, so care must be taken to find the most appropriate terms in *each database*.

### 2.1.2.1    Subject Headings

CAPLUS has over 200,000 Subject Index Headings, and these cover virtually all the sciences. Additionally there are a greater number of headings for organisms/species. It is helpful, particularly for the purpose of understanding what may be included in a 'concept' (Section 3.3), to be aware of some general aspects of Index Headings, including:

- Index Headings may occur either in the singular form (e.g. Cytoplasm) or in the plural form (e.g. Inhibitors), which may have implications for the SciFinder **Explore References: Research Topic** algorithms;

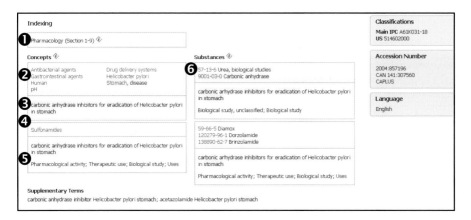

**Figure 2.1**    *Indexing of a record in CAPLUS in which different types of indexing are indicated. 1. CA Section Code; 2. Subject Index Headings; 3. Text-modifying phrase; 4. Substance Class Heading; 5. CAS Roles; 6. CAS Registry Numbers. SciFinder® screens are reproduced with permission of Chemical Abstracts Service (CAS), a division of the American Chemical Society*

- Substance classes that relate to pharmaceutical properties usually have the prefix 'anti' before the main term (e.g. Anticholesterolemics, Anti-AIDS Agents; see Figure 2.2);
- Substance Class Headings are generally in the plural form (e.g. Alkenes, Porphyrins, Sulfonamides);
- Subject Headings are generally in natural language order, e.g. Molecular cloning in CAPLUS, but this does not apply necessarily to MEDLINE, where the corresponding Index Heading is Cloning, molecular.

Subject Headings in CAPLUS are arranged in a hierarchy. An example of the hierarchy is given in Figure 2.3 for the Index Heading Helicobacter pylori.

It is not necessary for SciFinder users to understand the details of this indexing and indeed the hierarchy is not displayable, but it is helpful to understand some basic points since they are often incorporated into SciFinder search algorithms. This display shows:

- Helicobacter pylori is an Index Heading in CAPLUS, and there are 12,101 records in CAPLUS that have this heading (i.e. clicking on this Index Heading, Section 3.6, gives 12,101 answers);
- Helicobacter pylori is in an index hierarchy with five broader terms (indicated at different Broader Term levels in Figure 2.3) and one narrower term (NT1 = Narrow Term at the first level);
- The Index Heading has been used from 1997 onwards;
- The older Index Heading is Campylobacter pyloridis (1209 records in the database have this heading);
- The Index Heading is used (UF = Used For) whenever the original article refers to Campylobacter pylori or Helicobacterium pylori (and these terms may be included in the SciFinder search algorithm – Section 3.3.3);
- A list of chemical substances that are commonly reported in records which contain the Index Heading.

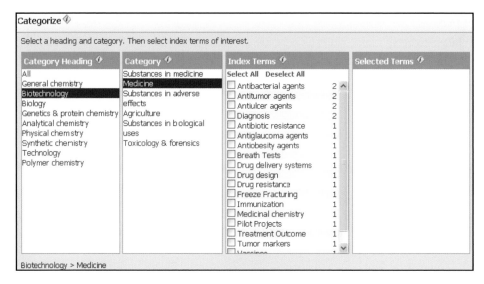

**Figure 2.2**    *Index Terms under Category Heading Biotechnology for the answer set described in Chapter 1 (carbonic anhydrase search). Note that the prefix 'anti' is commonly used for classes of drugs for specific treatments. SciFinder® screens are reproduced with permission of Chemical Abstracts Service (CAS), a division of the American Chemical Society*

```
   476   BT5    Organisms
 31447   BT4      Eubacteria
         BT3        Unassigned bacteria (non-CA heading)
  4889   BT2          Gram-negative bacteria
   559   BT1            Helicobacter
 12101   Helicobacter pylori
                 HNTE Valid heading during volume 126 (1997) to present.
  1209           OLD Campylobacter pyloridis
                   UF Campylobacter pylori
                   UF Heliobacterium pylori
     4           NT1 Helicobacter pylori pylori
                 RTCS Amoxicillin
                 RTCS Clarithromycin
                 RTCS Lansoprazole
                 RTCS Metronidazole
                 RTCS Omeprazole
                 RTCS Tinidazole
                 RTCS Urease
```

**Figure 2.3**    *Index hierarchy for Helicobacter pylori in CAPLUS. BT, Broader Term; NT, Narrower Term; HNOTE, Historical Note; UF, Used For; RTCS, Related Term Chemical Substance. The numbers on the left indicate the number of records in the database with the term indicated. Copyright© 2009 American Chemical Society (ACS). All Rights Reserved. Reprinted with permission*

In turn some comments on the display are:

- Index Headings are entered at the most precise level possible, so the Index Heading Helicobacter is used when only the class of organisms is reported; if Helicobacter pylori is specifically mentioned then its precise Index Heading is entered in CAPLUS. Accordingly, a search on the Index Heading Helicobacter will not retrieve narrower Index Headings;
- Index Headings may change over time. There is no way of knowing this from the display of **Analyze: Index Term** (e.g. in Figure 1.10), so the user needs to go through the list of index terms carefully and choose all that are appropriate.

---

*SciFinder Tip*

The simplest way to learn about subject indexing in SciFinder is to obtain an initial answer set and then to **Analyze: Index Term** or to **Categorize**. Subject indexing under **Categorize** is primarily found through Category Topics (in the second column of the Categorize display) (see Figure 2.2).

Users need to go down the list, i.e. '**Show More'**, partly because Index Headings may change with time. For example, there may be relatively few occurrences for a recently added Index Heading; because it is recent and because Index Headings are very precise, the really important Index Headings may be towards the end of **Analyze** lists, which are sorted by frequency. At times it may be easier to display **Analyze** lists in alphabetical order.

---

### 2.1.2.2   Substance Class Headings

Substance Class Headings are entered when the original article describes something new about the substance class or about a number of specific substances that are described by this class. For example, if reactions of a variety of individual alkenes are mentioned in the original text then the Substance Class Heading Alkenes is indexed.

### 2.1.2.3   Substances

CAS Registry Numbers underpin the CAS Registry System, and the only authorative source of CAS Registry Numbers is the REGISTRY database, which may be searched in SciFinder in a number of ways (Chapters 4 and 5). Each unique substance is given a separate CAS Registry Number and when something new is reported on the substance in the original article then the CAS Registry Number is added as an index term in the CAPLUS record. Not every substance in the original article is indexed and some CAS indexing policies relating to indexing of substances are indicated in Table 2.1.

CAS Registry Numbers are precise and comprehensive search terms for substances, and CAS Registry Numbers should nearly always be included in the search for substances.    Results from name-based searches (**Explore References:   Research Topic**) should be checked to see that the CAS Registry Number has been used (see Section 3.3.5).

**Table 2.1**  *Some examples of indexing policies for substances in CAPLUS*

| Indexing Policy | Note | Implications |
|---|---|---|
| The CAS Registry Number is indexed only if something new is reported for the substance in the original article. | Particularly in the introduction section, many original articles may summarize previously known information, and names or even structures of substances may appear. However, the CAS Registry Number is not indexed unless something *new* is reported for the substance. | If the focus is on comprehension it may be advisable to search for some common names of substances as well as for the CAS Registry Numbers. |
| For patents, CAS Registry Numbers are entered in CAPLUS mainly for substances that have been characterized (e.g. in the Experimental Section) or which are specifically mentioned in the Claims. | Generic (Markush) structures from patents are not indexed in CAPLUS (but specific examples are indexed). Substances mentioned in the discussion section (and for which no new information is reported) are not indexed. | If the generic structure describes a halogen (X) and if the specific example in the patent contains a bromine (Br), then the bromo compound will be indexed. Substructure searches should start with more general queries (e.g. with halogen (X) or with no substitution at the position involved). |
| The substance is indexed as precisely as possible. | The CAS Registry Number for morphine is listed if the original article refers to morphine, but the CAS Registry Number for morphine sulfate is listed if the article refers to the sulfate. Similarly, the CAS Registry Number for potassium is listed if the article refers to potassium (e.g. 'potassium levels in blood'), but for potassium ion if the article refers to K$^+$. | In some cases a few CAS Registry Numbers (e.g. the parent base and all its salts) may be needed to cover a 'substance'. |
| Since 1985, if a simple name for the substance is given by the author it may be included after the CAS Registry Number; CAS does not apply systematic nomenclature. | 'Diamox' is added after the CAS Registry Number 59-66-5 in Figure 2.1. However, in pre-1985 records, and in subsequent records where simple names were not in the original documents, CAS Registry Numbers alone appear. (For an example of the latter, see CAS Registry Number 9068-38-6 in Figure 3.9.) | CAS Registry Numbers should nearly always be used in search terms for substances. |
| If the original report refers to a number of substances of a certain class, then the substance class heading is also indexed. | The substance class heading 'sulfonamides' is used in Figure 2.1. | General information on substance classes may be searched effectively through the substance class index headings. |

There are five CAS Registry Numbers in Figure 2.1 and following each is a simple name. Often the name entered is the name used by the authors in the original document, but the inclusion of the name here is intended only to help with quick identification of the substance; these names are not systematic and cannot be relied upon for comprehensive searches, particularly since in many instances CAS Registry Numbers *without any names* appear in many records in CAPLUS (e.g. see Figure 3.9).

### 2.1.2.4 CAS Roles

All CAS Registry Numbers and Substance Class Headings are followed by CAS Roles, which are index terms that relate to the actual research on the substance in the original document. CAS Roles have been assigned by the indexers since October 1994, while CAS Roles in the records in CAPLUS prior to that time have been assigned 'algorithmically' (i.e. the CAS Roles have been assigned by computer-based searches which involved combinations of searches in section codes, controlled terms, and keywords).

CAS Roles (Appendix 2) relate at a higher level to general properties such as agricultural, chemical, biological, environmental, and medical properties, and to preparations and reactions. Within these properties there are even more specific roles. For example, the role Preparation is further divided, among other things, into bioindustrial and synthetic preparations.

---

*SciFinder Tip*

CAS Roles are shown in displays in records (e.g. see Figure 2.1). They are also used to assist with specification of the Category (see Column 2 in Figure 2.2 where 'Substances in adverse effects' is listed).

Searching with CAS Roles is a precision tool, which is very useful for focusing on specific studies, particularly when large numbers of records occur for individual substances.

---

### 2.1.2.5 Text-Modifying Phrases

Text-modifying phrases are terms that follow Index Headings or CAS Registry Numbers. They are often terms from the original article that relate most directly to the Index Heading, so may be considered as author-related terms that qualify the Index Heading. The inclusion of these terms is significant since one of the features of SciFinder is that the user may choose answers in which concepts searched are 'closely associated'. In general SciFinder defines terms to be 'closely associated' when they appear in the title, in a single sentence in the abstract, or within a single index term and its text-modifying phrase. The assumption is that terms that are 'closely associated' (rather than anywhere in the reference) are more directly related, so the inclusion of text-modifying phrases after the Index Heading provides an important level of precision in the choice of answer sets.

In cases where text-modifying phrases are common to a number of Index Headings, the headings are grouped and the single text-modifying phrase is applied. Therefore in Figure 1.5 the text-modifying phrase 'roles of $\alpha$ and $\beta$ carbonic anhydrases of Helicobacter pylori in urease-dependent response to acidity and in colonization of murine gastric mucosa' is closely associated with *each* of the six Index Headings before it.

---

*SciFinder Note*

An Index Heading and its text-modifying phrase are considered 'closely associated' in SciFinder. However, when several Index Headings have a common text-modifying phrase, the *headings* are not 'closely associated' with each other.

---

### 2.1.2.6   Subheadings

The first part of the text-modifying phrase may contain a subheading, which is then followed by a semi-colon. For example, in Figure 1.5 the entries 'mucosa' and 'periplasm' are Subheadings to Stomach and Organelle respectively. Users see Subheadings mainly in histograms that display the analysis of index terms (Figure 2.4). For example, there are two entries shown under the heading Stomach:

- Stomach with subheading 'disease' appears in eight records;
- Stomach with subheading 'neoplasm' appears in six records.

The user may choose either or both of these terms – it depends on the search requirements.

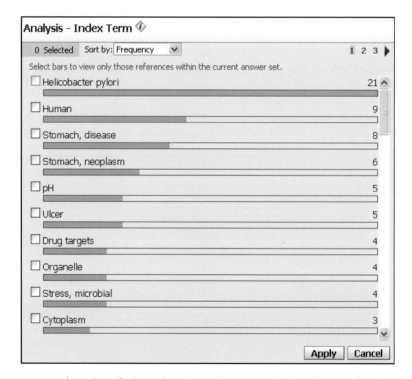

**Figure 2.4**   *Display of **Analysis: Index Term** showing index headings and subheadings in CAPLUS for search on 'carbonic anhydrase with helicobacter pylori'. SciFinder® screens are reproduced with permission of Chemical Abstracts Service (CAS), a division of the American Chemical Society*

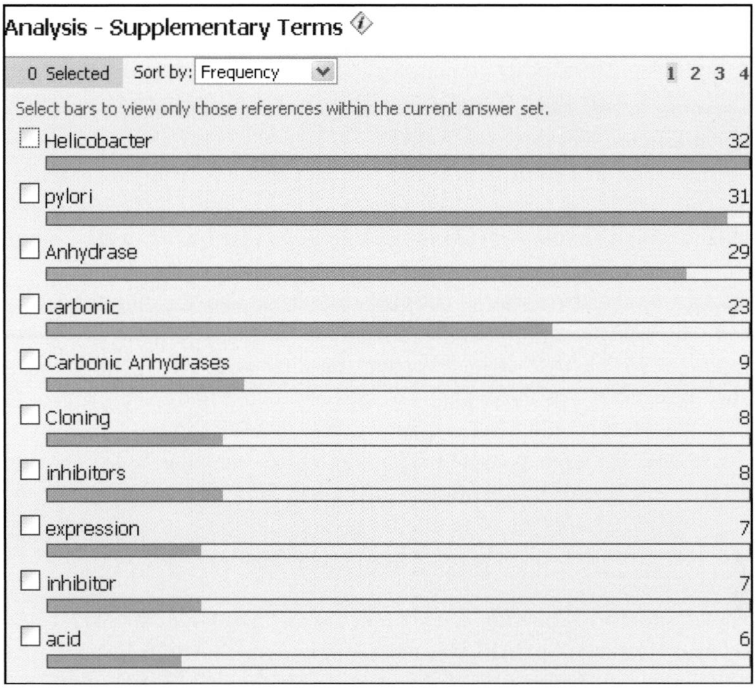

**Figure 2.5** *Display of* **Analysis: Supplementary Terms** *for search on 'carbonic anhydrase with Helicobacter pylori'. SciFinder® screens are reproduced with permission of Chemical Abstracts Service (CAS), a division of the American Chemical Society*

### 2.1.2.7   Supplementary Terms

Supplementary terms are natural language words or phrases that are entered to provide key additional information about the content of the document. Generally they relate to authors' terminology, and while they are not controlled vocabulary (systematic) terms, they provide another option to refine answer sets.

In a manner similar to that in which text-modifying phrases are often author-related terms that expand on the index term, Supplementary Terms may be considered as index-related terms that expand on the terms used by authors in titles.

A display of results from **Analyze: Supplementary Term** for the query in Section 1.2.1 is shown in Figure 2.5, and it is noted that Helicobacter and pylori are listed separately. This is because most words in the Supplementary Term field usually are considered separately in the **Analysis** list.

## 2.2   NLM Bibliographic Database (MEDLINE)

The MEDLINE database contains more than 18 million references to journal articles in the biomedical sciences. It covers more than 5200 journals published from the early 1950s (see Appendix 1 for links) and is updated four times a week. A Fact

Sheet which summarizes the content of the database is available (see Appendix 1 for links).

An example of the content of a single record is shown in Figure 1.7. After the title and bibliographic information, the majority of records contain the abstract as presented in the original article. The indexing follows.

A key to indexing in MEDLINE is the Medical Subject Headings (MeSH®) which are outlined in another Fact Sheet (see Appendix 1 for links). There are over 19,000 main headings, which are constructed into a thesaurus that links broader, narrower, and related terms in the hierarchy.

The structure of the thesaurus is detailed elsewhere and the full thesaurus may be downloaded (see Appendix 1 for links). These are very useful links and at some stage frequent users of MEDLINE should become familiar with MeSH. However, a quick appreciation of what is involved may be seen through Figure 2.6, which shows the thesaurus for the main heading Helicobacter pylori in MEDLINE.

In this case MEDLINE has two broader hierarchies and the NOTE is informative. MEDLINE indexing also has MN numbers (Medical Subject Heading), historical terms, and Used For terms. The series of two letter codes after the term AQ (Allowable Qualifier) are abbreviations for specific branches of investigation such as CH = Chemistry and ME = Metabolism.

In a manner similar to the way CAS Roles qualify the CAS Registry Numbers and Substance Class Headings in CAPLUS, most index terms in MEDLINE are qualified. A full list of allowable qualifiers is available (see Appendix 1 for links), although to assist users the qualifiers and the acronym are always present in the actual database record.

```
      0    BT5      B Organisms
  77283    BT4      Bacteria
    728       BT3      Proteobacteria
     86          BT2      Epsilonproteobacteria
      0       BT4    B Organisms
  77283       BT3      Bacteria
  11049          BT2      Gram-Negative Bacteria
    732             BT1   Helicobacter
  22126             Helicobacter pylori
  22126             MN    B3.440.500.550.
  22126             MN    B3.660.150.280.550.
             DC    an INDEX MEDICUS major descriptor
  NOTE  A spiral bacterium active as a human gastric pathogen. It is a gram-negative, urease-positive,
  curved or slightly spiral organism initially isolated in 1982 from patients with lesions of gastritis or peptic
  ulcers in Western Australia. Helicobacter pylori was originally classified in the genus
  CAMPYLOBACTER, but RNA sequencing, cellular fatty acid profiles, growth patterns, and other
  taxonomic characteristics indicate that the micro-organism should be included in the genus
  HELICOBACTER. It has been officially transferred to Helicobacter gen. nov. (see Int J Syst Bacteriol
  1989 Oct;39(4):297–405).

             INDX infection: coord IM with HELICOBACTER INFECTIONS (IM)
             AQ  CH CL CY DE EN GDGE IMIP MEPH PY RE UL VI
             PNTE Campylobacter (1984-1990)
             HNTE 91
             MHTH NLM (1991)
      0    UF   Campylobacter pylori/CT
```

***Figure 2.6***  *Index hierarchy for Helicobacter pylori in MEDLINE. BT, Broader Term; NT, Narrower Term; MN, MeSH Number; AQ, Allowable Qualifier; PNOTE, Previous Indexing Note; HNOTE, Historical Note; MHTH, MH Thesaurus; UF, Used For*

For example, in Figure 1.7 the heading Bacterial Proteins is followed by the qualifier ME, metabolism.

In addition to the index headings and allowable qualifiers, MEDLINE adds an asterisk (*) to those index headings considered to be key terms related to the article. These asterisks may appear in **Analyze: Index Term** lists (see Figure 1.10).

Records in MEDLINE also may contain CAS Registry Numbers, chemical names, and chemical terms. While CAPLUS contains more than 60 million different CAS Registry Numbers, only slightly more than 57,500 of these have listings in MEDLINE. The implications of this are discussed in Chapter 6.

## 2.3   CAS Substance Database (REGISTRY)

The classification of substances in chemical substance databases generally follows the conventions used by chemical scientists, but the catch is that in some important cases it does not! The main reason is that while it may be easy to write down a structure or describe it in words, the somewhat loose descriptions used, and the complexities of structures themselves, may not easily be specified unambiguously in a digital database. Consequently it helps if searchers understand the index policies that have been applied. Some of these are discussed below, while more detail in given in Chapters 4 and 5, and some examples are given in Appendix 4.

Up-to-date information on REGISTRY is available (see Appendix 1 for links). As of January 2009 the database contains more than 42 million records for organic and inorganic substances and more than 60 million records for sequences. An example of a record in SciFinder is shown in Figure 1.6, but a more complete record (Figure 2.7) is discussed below for 'Carbonic anhydrase inhibitor 6063' (CAS Registry Number 59-66-5).

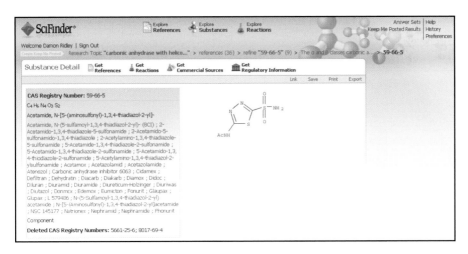

*Figure 2.7*   *Example of substance identifying information (CAS Registry Number, molecular formula, names, and structure) in records in REGISTRY. SciFinder® screens are reproduced with permission of Chemical Abstracts Service (CAS), a division of the American Chemical Society*

Records for substances in REGISTRY start with the CAS Registry Number, molecular formula, reported names, and chemical structure. This record also lists deleted CAS Registry Numbers, which may occur for a number of reasons including a change in policy for registration or a correction of the structure of the substance in the literature. However, deleted CAS Registry Numbers are of little consequence to SciFinder users since they are automatically included in searches when answers from REGISTRY are transferred to other databases.

Note that the term 'Component' appears after the names in this record. 'Component' indicates that the CAS Registry Number (here 59-66-5) appears as a component in a multicomponent substance. (For further information, see Appendix 4, Section A4.2.)

Next follows information on the types of documents and the CAS Roles relating to the substance in CAPLUS (Figure 2.8). It is helpful to browse this information to get some understanding of the types of records. For example, in this case there is information on analysis and on biological studies from both patent and nonpatent literature.

Mention also is made of 'Nonspecific Derivatives', which are substances very similar to, or derived from, the substance. Such derivatives may include products from combinatorial studies, products from reactions where the substance is treated in some way (e.g. sulfonated) but where the products are not fully characterized, or products resulting from modifications of polymers (Section 6.10).

Records for substances with specific structures in REGISTRY then contain extensive information on predicted properties. These are calculated using Advanced Chemistry Development Software and include predicted properties such as bioconcentration factors, Koc, logD, Mass Solubility, and pKa. Much of this information is of particular value to those in the broad field of drug research, and SciFinder offers a **Refine: Property Availability** option (Section 5.3.2) for including this information in searches. Figure 2.9 shows some examples of predicted properties for the substance in Figure 2.7.

REGISTRY records may also contain experimental property information (Figure 2.10), which is linked to the CAPLUS records, and spectral information, which may be linked to actual spectra (Figure 2.11). The total number of predicted and experimental property

Document Types: Conference, Dissertation, Journal, Patent, Report

| Role | Patents | Nonpatents | Nonspecific Derivatives from Patents | Nonspecific Derivatives from Nonpatents |
|---|---|---|---|---|
| Analytical study | ✓ | ✓ | | ✓ |
| Biological study | ✓ | ✓ | ✓ | ✓ |
| Formation, nonpreparative | | ✓ | | ✓ |
| Occurrence | | ✓ | | |
| Preparation | ✓ | ✓ | ✓ | ✓ |
| Process | ✓ | ✓ | | ✓ |
| Properties | ✓ | ✓ | ✓ | ✓ |
| Reactant or reagent | ✓ | ✓ | | ✓ |
| Uses | ✓ | ✓ | ✓ | ✓ |

**Figure 2.8** *Display in REGISTRY that shows the types of documents and CAS roles that are found for the substance (Figure 2.7) in CAPLUS. SciFinder® screens are reproduced with permission of Chemical Abstracts Service (CAS), a division of the American Chemical Society*

| Predicted Properties: | Biological | Chemical | Density | Lipinski and Related | Spectra | Structure-related | | |
|---|---|---|---|---|---|---|---|---|

| Biological Properties | Value | Conditions | Notes | Top |
|---|---|---|---|---|
| Bioconcentration Factor | 1.0 | pH 1 Temp: 25 °C | (26) | |
| Bioconcentration Factor | 1.0 | pH 2 Temp: 25 °C | (26) | |
| Bioconcentration Factor | 1.0 | pH 3 Temp: 25 °C | (26) | |
| Bioconcentration Factor | 1.0 | pH 4 Temp: 25 °C | (26) | |
| Bioconcentration Factor | 1.0 | pH 5 Temp: 25 °C | (26) | |
| Bioconcentration Factor | 1.0 | pH 6 Temp: 25 °C | (26) | |
| Bioconcentration Factor | 1.0 | pH 7 Temp: 25 °C | (26) | |
| Bioconcentration Factor | 1.0 | pH 8 Temp: 25 °C | (26) | |
| Bioconcentration Factor | 1.0 | pH 9 Temp: 25 °C | (26) | |
| Bioconcentration Factor | 1.0 | pH 10 Temp: 25 °C | (26) | |

| Chemical Properties | Value | Conditions | Notes | Top |
|---|---|---|---|---|
| Koc | 17.2 | pH 1 Temp: 25 °C | (26) | |
| Koc | 17.2 | pH 2 Temp: 25 °C | (26) | |
| Koc | 17.2 | pH 3 Temp: 25 °C | (26) | |

**Figure 2.9** *Example (for the substance in Figure 2.7) of predicted properties in REGISTRY records. Links to property groups are available from the top line. The list of predicated properties is extensive and only the first part is shown here. SciFinder® screens are reproduced with permission of Chemical Abstracts Service (CAS), a division of the American Chemical Society*

| Experimental Properties: | Biological | Chemical | Lipinski and Related | Spectra | Thermal | | |
|---|---|---|---|---|---|---|---|

| Biological Properties | Value | Conditions | Notes | Top |
|---|---|---|---|---|
| ADME (Absorption, Distribution, Metabolism, Excretion) | See full text | | (5) CAS | |
| Half-Life (Biological) | See full text | | (5) CAS | |

| Chemical Properties | Value | Conditions | Notes | Top |
|---|---|---|---|---|
| Acid/Base Dissociation Constant (Ka/Kb) | See full text | | (1) CAS | |
| Acid/Base Dissociation Constant (Ka/Kb) | See full text | | (2) CAS | |
| Acid/Base Dissociation Constant (Ka/Kb) | See full text | | (3) CAS | |
| Acid/Base Dissociation Constant (Ka/Kb) | See full text | | (4) CAS | |
| logP | See full text | | (4) CAS | |
| logP | See full text | | (9) CAS | |
| logP | See full text | | (10) CAS | |
| Solubility | See full text | | (24) CAS | |

| Lipinski and Related Properties | Value | Conditions | Notes | Top |
|---|---|---|---|---|
| logP | See full text | | (4) CAS | |
| logP | See full text | | (9) CAS | |

**Figure 2.10** *Example (for the substance in Figure 2.7) of experimental properties in REGISTRY. Links to property groups are available from the top line. The list of experimental properties is extensive and only the first part is shown here. SciFinder® screens are reproduced with permission of Chemical Abstracts Service (CAS), a division of the American Chemical Society*

information in SciFinder exceeds 2 billion, and SciFinder now contains more than 23.8 million predicted proton NMR spectra.

Note that above the record (Figure 2.7) is a row starting with Substance Detail and followed by **Get References, Get Reactions, Get Commercial Sources**, and **Get Regulatory Information**. These provide direct links to information on the substance in CAPLUS/MEDLINE, CASREACT, CHEMCATS, and CHEMLIST respectively, and offer one way in SciFinder through which information of one type (e.g. substance information) can be transferred easily from one database to another database where information of another type (e.g. commercial suppliers for the substance) is available.

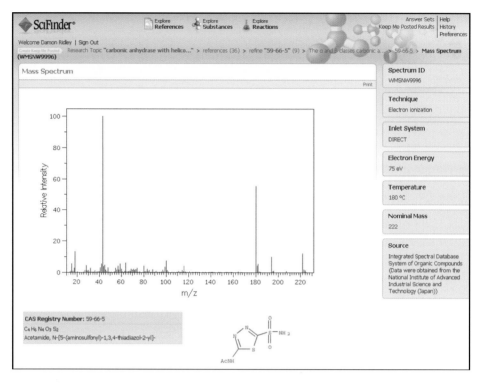

**Figure 2.11**    *Example of a Mass Spectrum in REGISTRY. SciFinder*® *screens are reproduced with permission of Chemical Abstracts Service (CAS), a division of the American Chemical Society*

## 2.4   CAS Chemical Reaction Database (CASREACT®)

The reaction database contains around 16 million single-step and multistep reactions selected mainly from journal articles since around 1974 (see Appendix 1 for links), although reactions reported back to 1840 and more recently extensive chemical reaction information from patents have been added. A typical example in SciFinder is shown in Figure 2.12. In the reaction database, all atoms and bonds are correlated between the starting material and product, and bonds being formed or broken are tagged. This enables precise reactions to be retrieved easily, which is important since a chemist may wish to know not only that a substance has been prepared but also about preparations involving formation of specific bonds.

It is helpful to appreciate that only key or representative new reactions from original documents may be fully indexed. Even so, it is one of the world's premier reaction databases and generally produces many more relevant answers than alternative reaction databases. SciFinder offers a number of other options for searching for chemical reactions, and further details are discussed in Chapter 7.

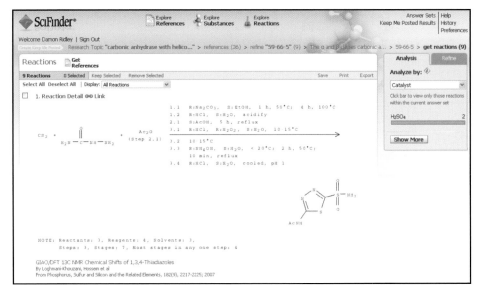

**Figure 2.12**  *Example of a record in the reaction database CASREACT. Different* **Analysis** *and* **Refine** *options appear (by default* **Analyze: Catalyst** *is shown). SciFinder® screens are reproduced with permission of Chemical Abstracts Service (CAS), a division of the American Chemical Society*

## 2.5  CAS Chemical Catalog Database (CHEMCATS®)

As of January 2009, the CAS chemical catalog database contains information on more than 29 million commercially available substances from around 900 suppliers and 1000 catalogs (see Appendix 1 for links). Each record contains the catalog information for the substance (e.g. chemical and trade names, the company names, and addresses), as well as supplier information (e.g. pricing terms).

It is neither possible nor necessary to search directly in CHEMCATS. Instead the strategy is to find the substance in REGISTRY first. When a substance appears in one of the chemical catalogs, a link (**Get Commercial Sources**) is provided to the chemical catalog database and it is a simple matter for the user to scroll through the company information to identify a local supplier.

## 2.6  CAS Regulatory Information Database (CHEMLIST®)

In order for national authorities to have a mechanism to regulate trade in chemicals, many countries require companies to register substances prior to their manufacture or distribution. Such inventories are essential, for example, for the monitoring of illegal substances and for keeping track of environmental issues relating to chemicals. The primary reference point in these national inventories is the CAS Registry Number.

CAS has built a database of regulated chemical substances from a number of national and international chemical inventories and regulatory lists. The database contains over 240,000 chemical substances and is updated weekly. Details of the content are available (see Appendix 1 for links) and a typical record is shown in Figure 2.13.

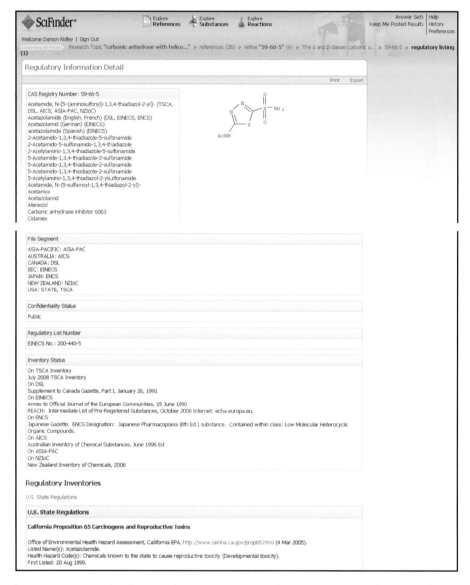

**Figure 2.13**   *Example of a record in CHEMLIST. Regulatory inventories where the substance is registered are indicated. Only part of the record is shown here. SciFinder® screens are reproduced with permission of Chemical Abstracts Service (CAS), a division of the American Chemical Society*

To access CHEMLIST data the user needs first to find the chemical substance in REGISTRY and then to click on the link labelled **Get Regulatory Information**.

---

*SciFinder Tip*

Links from REGISTRY to the other SciFinder databases may be obtained from the full substance record (Figure 2.7). Links are also available from *abbreviated* displays of substances, as shown in Figure 2.14. Such displays appear automatically after a substance search has been performed or after **Get Substances** from a bibliographic answer set is chosen.

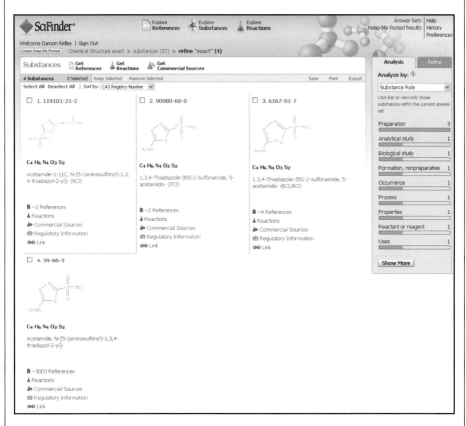

***Figure 2.14*** *Abbreviated substance display in SciFinder. Screens of this type appear for initial answers under **Explore Substances** and after **Get Substances** is chosen from bibliographic records. SciFinder® screens are reproduced with permission of Chemical Abstracts Service (CAS), a division of the American Chemical Society*

---

## 2.7   Summary of Key Points

- The two bibliographic databases in SciFinder are CAPLUS and MEDLINE, and together they contain more than 48 million records;
  - Some of the records in each database will be derived from the same original article, so the number of original articles covered will be less – perhaps around 36 million;
  - When each database has a record for the same original article, the records will be different and this in particular applies to the indexing;
  - In any event, together the databases cover a vast amount of information in the sciences.
- Both CAPLUS and MEDLINE have extensive indexing (through the CA Lexicon® and MeSH respectively). SciFinder uses the indexing 'behind the scenes' in algorithms under **Explore References: Research Topic** and SciFinder displays indexing in actual records and through **Analyze: Index Term** and **Categorize**;
- The main substance database, REGISTRY, offers the world's largest coverage of chemical substances and sequences in the one location. From records in this database may be found:
  - references to information on the substances in CAPLUS and in MEDLINE;
  - commercial sources of substances in the substance database CHEMCATS;
  - information on the registration of substances in the substance database CHEMLIST;
  - predicted and experimental properties, and spectra.
- CASREACT is a specialist chemical reaction database with extensive information on reactions from journals and patents;
- It is a simple matter on SciFinder to transfer information from one database to another.

# 3

# Explore by Research Topic

## 3.1 Introduction

Information on topics is obtained from the CAPLUS and MEDLINE databases. After signing in to SciFinder, the **Explore References: Research Topic** screen appears and the query is entered. A list of topic candidates is obtained, the required boxes are chosen, and the initial answer set is displayed. The steps and parts of initial SciFinder screens are shown in Figure 3.1.

None of this requires any knowledge of how the search was performed, or of the databases, and often good answers are obtained immediately with the minimum of effort. SciFinder is very easy to use at the basic level, and scientists will know that some of the best results in science come from the simplest experiments. Therefore any searcher who is unsure of how to approach a problem should just try something!

However, how does SciFinder convert the initial query to concepts, i.e. what is 'behind the scenes' in going from Step 1 to Step 2? What has been searched to give the numbers of candidates? What are the issues in the choice of the appropriate candidates from the list and which features from the initial answer screen should be used and when? Perhaps most importantly, how may an understanding of these issues help to increase comprehensiveness and precision in answers and to assist in the production of creative solutions to complicated problems?

This chapter addresses these questions.

## 3.2 How SciFinder Converts the Query to a List of Candidates

After **Search** is clicked (Step 1, Figure 3.1), SciFinder automatically applies a number of algorithms to the search query and the subsequent list of candidates may contain the following terms:

- 'As entered' contains the actual words in the order entered in the query, although some variation in words not searched (e.g. prepositions and stop-words) may be allowed;

*Information Retrieval: SciFinder®, Second Edition* Damon D. Ridley
© 2009 John Wiley & Sons, Ltd

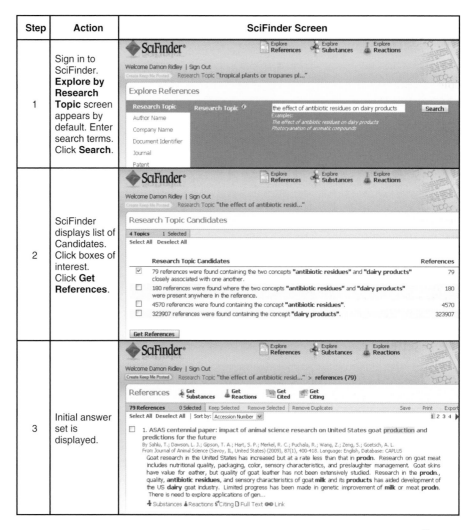

***Figure 3.1*** *The steps involved with **Explore References: Research Topic***. *SciFinder*® *screens are reproduced with permission of Chemical Abstracts Service (CAS), a division of the American Chemical Society*

- 'Concept' includes the term entered and synonyms that have been identified by SciFinder algorithms;
- 'Closely associated' indicates the concepts that appear near each other in the record (see Section 3.2.1 for a more precise definition);
- 'Anywhere in the reference' indicates the concepts that appear in the record.

SciFinder identifies individual concepts through the prepositions, conjunctions, and stop-words entered in the query. While the actual preposition (e.g. of, with, at) is of no significance, the conjunction chosen (and, not, or) may be critical. Certain stop-words such as 'the effect', 'information on', and so forth are also used to identify concepts and may not be searched. Stop-words are recognized as they are not included within the 'concepts' shown in the list of candidates. For example, **Explore References: Research Topic** 'the effect of antibiotic residues on dairy products' gives the candidate list shown in Step 1, Figure 3.1. 'The effect' is not listed, and indeed any requirement that it be present would dramatically reduce the answers obtained because authors would be unlikely to use such a term consistently.

---

*SciFinder Tip*

Examine the list of candidates and check if a term in the query is not listed. If not, then consider whether the term is needed. It is unlikely that it will be, but if needed then consider searching a short phrase (e.g. 'effect antibiotic' in the example above) and the stop-word may now be searched.

---

### 3.2.1　Search Fields

Next SciFinder searches for the 'concepts' in the title, abstract, and index fields, and may produce Research Topic Candidates where all the concepts are 'closely associated' and 'anywhere in the reference'. The entry 'closely associated' indicates that the terms in the concept are in the title, in the same sentence in the abstract, or in a single Index Heading and its text-modifying phrase, while 'anywhere in the reference' means that the terms are present somewhere in the title, abstract, and index fields (effectively Boolean 'AND' has been applied). The usual assumption is that the closer terms are then the more directly they are related, so the former answer set may afford a greater level of precision.

---

*SciFinder Tip*

Records in MEDLINE do not have text-modifying phrases associated with Index Headings, so the choice of 'closely associated' terms may be too restrictive if comprehensive answer sets in MEDLINE are important.

Also remember that each MEDLINE Index Heading is in a separate 'sentence', that is the headings are *not* 'closely associated'.

---

### 3.2.2　Candidates

SciFinder then displays candidates where combinations of *some* of the concepts are 'closely associated' or 'anywhere in the reference', and finally displays candidates with a number of answers for the individual concepts (Figure 3.2).

The number of candidates displayed depends on the number of concepts identified, and whether Boolean operators have been included in the question. Some examples are shown in Table 3.1.

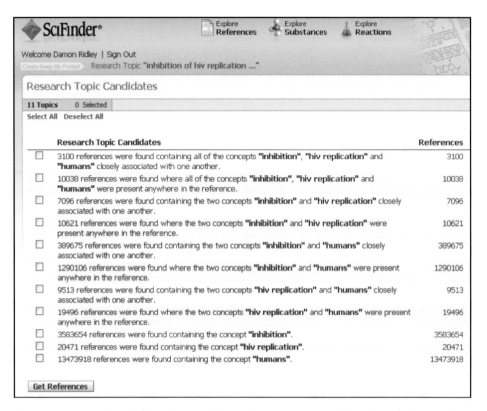

**Figure 3.2**   *List of candidates from **Explore References: Research Topic** 'inhibition of HIV replication in humans'. Three concepts are identified, and numbers of references with all three concepts followed by combinations of two of the concepts and then the individual concepts are listed. SciFinder*® *screens are reproduced with permission of Chemical Abstracts Service (CAS), a division of the American Chemical Society*

---

*Why Does SciFinder Break up Queries into Concepts and Then Show Various Combinations of the Concepts?*

SciFinder provides choices from precise options at the top of the candidate list to more general options at the bottom, and the user may immediately get a snapshot of the types of answers for the specific query. A search with too many concepts may be too restrictive and a search with fewer concepts may provide a better initial answer set. SciFinder provides actual numbers of records for each candidate to assist the user.

However, users should not be deterred with initial answer sets of a few thousand records since SciFinder has many functions to narrow answers in systematic ways.

---

The first task is to look carefully at the concepts that have been identified in order to verify that the algorithm has been applied in the intended manner. In the example chosen (Figure 3.2 and Entry 3 in Table 3.1) the concepts identified are 'inhibition', 'HIV

**Table 3.1**  *Examples of number of candidates identified from different **Explore References: Research Topic** entries*

| Entry | Explore References: Research Topic | Number of concepts | The concepts identified | | Number of candidates usually listed |
|---|---|---|---|---|---|
| 1 | Inhibition of replication of hiv in humans | 4 | 'inhibition' 'replication' | 'hiv' 'humans' | 26 |
| 2 | Inhibition and replication and hiv and humans | 4 | 'inhibition' 'replication' | 'hiv' 'humans' | 6 |
| 3 | Inhibition of hiv replication in humans | 3 | 'inhibition' 'hiv replication' | 'humans' | 11 |
| 4 | Inhibition and hiv replication and humans | 3 | 'inhibition' 'hiv replication' | 'humans' | 5 |
| 5 | Inhibition of human hiv replication | 2 | 'inhibition' 'human hiv replication' | | 4 |
| 6 | Inhibition hiv replication humans | 2 | 'inhibition hiv' 'replication humans' | | 3 |

replication', and 'humans', and the result is a list of 11 candidates where the concepts are variously listed as 'closely associated' or 'anywhere in the reference'.

However, if the question is asked in a slightly different way, other concepts may be identified and different numbers of candidates may be listed (Table 3.1). For example, because of the extra preposition in Entry 1 (Table 3.1), four concepts are identified and they, and their various combinations, are listed as 'closely associated' or 'anywhere in the reference' to give 26 candidates. On the other hand, Entry 2 (Table 3.1) in which the terms are connected with 'and' has only 6 candidates. The essential difference is that use of 'and' here is strictly interpreted so that terms are identified only as 'anywhere in the reference' and the 'closely associated' candidates are not displayed. Other entries in Table 3.1 are similarly explained, although if four terms are entered consecutively (Entry 6, Table 3.1) SciFinder automatically breaks them into two concepts. It is not advisable to search for several words in a single concept and so SciFinder guides the user through breaking up several words (which are grouped together) into different concepts.

Conjunctions 'and, not, or' may be entered in the query, but again the list of candidates needs to be studied carefully. Thus, in recognition that searchers may at times use 'and' when 'or' is intended, SciFinder sometimes interprets the conjunction 'and' in the more general sense (i.e. as 'or'). It all depends on the actual query. For example, while many of the candidates listed after **Explore References: Research Topic** 'treatment of hiv in men and women' are similar to the candidates listed after the entry 'treatment of hiv in men or women' (i.e. 'and' is interpreted in part as 'or' – however, it is better to use 'or' rather than 'and' for synonyms); nevertheless there are some unique candidates in each set.

At all times care must be exercised in the use of the conjunction 'not' since relevant records may be excluded. Indeed, in general, use of the word 'not' is not advised, and it is preferable to use some of the alternative strategies mentioned in this chapter to make answer sets more precise.

---

*SciFinder Tip*

Carefully consider which terms to include in concepts and where to place prepositions between the terms.

  Avoid the use of Boolean AND and NOT. In the former case enter instead a preposition otherwise, if appropriate, Boolean OR. For example, avoid using 'men and women' (which requires both words in the records retrieved) if the requirement really is 'men or women'.

---

### 3.2.3   Notes on Terms Entered

*3.2.3.1   Number of Concepts*

At times users enter too many terms, and more appropriate answer sets may be retrieved when fewer terms are employed. While all the concepts may be in the original publication, the records being searched include only title, abstract, and index entries in which all the terms may not be mentioned. Through the display of combinations of the concepts, SciFinder is guiding the searcher who, for example, in cases where very few records mention all the concepts may instead be alerted to and choose candidates with fewer concepts before exploring the question further.

*3.2.3.2   Prepositions*

Inclusion of more prepositions (Entry 1, Table 3.1) produces the greatest number of candidates, and these candidates will ultimately include the options displayed in Entries 2 to 6 (Table 3.1). Having more options is helpful since the user has a better idea beforehand of the types of answer sets that may be of interest. However, if more than four concepts are identified the candidate list may be tedious to work through. For example, the permutations and combinations of five concepts linked with prepositions gives 57 candidates.

*3.2.3.3   Records for Individual Concepts*

The number of records for individual concepts at the bottom of the list provides an important piece of information. For example, **Explore References: Research Topic** 'removal of cyanide from wastewater from gold mine tailings' produces only 5 and 13 records where all concepts are 'closely associated' and 'anywhere in the reference' respectively. This is a very small number of hits for such a topic in a database that is particularly strong in the area of mining and ore extraction. The problem is evident when the number of hits for the individual concepts is examined, and in this case there are just over 450 hits for 'gold mine tailings'. While this may be an acceptable number for the individual concept, it is too restrictive when combined with the other concepts.

  The issue is that when two or three words are within the same concept, SciFinder will look for records containing *all* of the words in the same sentence, but there may be sentences that have just one of the words or the word with quite different second or third words. So while there are just over 450 hits for the concept 'gold mine tailings', there are around 6000 hits for the concept 'gold mine' and over 340,000 for the concept 'gold'.

This last answer set includes words like 'gold mine wastes' and 'gold processing plants' that are clearly relevant to questions that concern removal of cyanide from wastewater in the manufacture of gold. Accordingly **Explore References: Research Topic** 'removal of cyanide from wastewater with gold' is better, and now around 100 references are retrieved with all four concepts 'closely associated'.

---

*SciFinder Tip*

It is preferable to enter between two to five concepts in the initial query, to avoid too many words within a single concept and to avoid words that are redundant. SciFinder suggests alternatives, and indeed the ability of SciFinder to guide the searcher through alerting of alternatives greatly facilitates searches on topics.

---

### 3.2.3.4   Number of Terms in Concepts

If more than three terms are initially detected in a concept, SciFinder may divide this concept further and an example is shown in Entry 6 (Table 3.1). If the working of this algorithm causes difficulties then it is a simple matter to interpose a preposition, or else restrict the consecutive terms to fewer than four words. For example, **Explore References: Research Topic** 'fourier transform infrared spectroscopy' will produce two concepts ('fourier transform' and 'infrared spectroscopy'), whereas 'fourier transform infrared' produces a single concept that effectively has the same outcome in a search.

### 3.2.3.5   Distributed Modifiers

At times the concepts identified will not be as intended, in which case alternative entries need to be made. A common problem lies with the issue of 'distributed modifiers', i.e. different terms that qualify another term. For example, it is perfectly acceptable to express in the English language 'I want information on liver or kidney diseases', but since SciFinder uses 'or' to identify concepts, the concepts identified in this query are 'liver' and 'kidney diseases'. The appropriate entry may thus be 'liver diseases or kidney diseases', where SciFinder identifies the concepts 'liver diseases' and 'kidney diseases' as required.

Depending upon the original entry, there may be a few other reasons why the algorithm did not interpret the query as intended. It is a simple matter to note the concepts SciFinder identifies and to make logical revisions to the original query where necessary.

## 3.3   How Is a Concept Derived?

The list of candidates mentions 'concepts', which are determined by SciFinder after application of a number of rules.

### 3.3.1   Automatic Truncation

Truncation of terms allows for retrieval of words that contain a common word fragment, and SciFinder applies truncation automatically. While this saves the user from having to think about truncation, exactly where to truncate is a tricky problem for any algorithm.

Truncation too late may exclude relevant terms, whereas truncation too early may retrieve irrelevant terms that happen to contain the word fragment.

In order to determine how the automatic application of truncation has been applied it is necessary to look at the 'hit' terms in the full records. Hit terms are terms in records that caused the answer to be retrieved, and in SciFinder hit terms appear with greyed backgrounds. For example, it is found that within the concept 'inhibition' SciFinder retrieves 'inhibitor' as well as 'inhibiting', 'inhibit', 'inhibitors', and so forth.

Automatic truncation works very effectively in the great majority of cases. However, if the answers include references where truncation appears not optimal, it may be better to work through answers manually or to use analysis or refinement options (Section 3.3.5 and following sections) rather than make decisions at this stage.

Truncation is a complex issue. The algorithm in SciFinder favours comprehension over precision, since it is assumed it is better to allow *users* to make decisions on relevancy of answers. It is better to know what is present, rather than not to know what may have been missed. Solutions to issues relating to application of automatic truncation are discussed later.

### 3.3.2   Singulars, Plurals, Tenses (Past, Present, Future)

In most cases, SciFinder searches for singular and plural forms of nouns when just one form is entered, and various tenses of verbs are searched automatically also. An exception is when SciFinder finds that the term entered matches an Index Heading in CAPLUS or in MEDLINE or the exact name of a substance in REGISTRY; in such cases the alternative singular or plural may not be applied automatically. For example, **Explore References: Research Topic** 'carbonic anhydrase with Helicobacter pylori' retrieves the record in CAPLUS in Figure 1.5, but does not retrieve the record in MEDLINE in Figure 1.7. The latter record is retrieved with **Explore References: Research Topic** 'carbonic anhydrases with Helicobacter pylori'. The issue relates to matching the different Index Headings in CAPLUS and in MEDLINE.

---

*Why Aren't Singulars/Plurals Automatically Searched on* All *Occasions?*

The issue is to write algorithms for all possible search entries (users think differently!) that balance comprehension/precision in answers from databases with billions of searchable words. SciFinder may draw the line when exact matches with Index Headings are found, since it is considered that Index Headings themselves provide a good balance and that the inclusion of additional singulars/plurals may compromise the search outcome.

The solution is to be aware of the issue, to look at hit terms in answers and to add singulars/plurals in the query if this best meets the requirements for the search at hand.

---

There are two ways to check outcomes: look through highlighted terms in records or start a new search with just the terms in question and check the number of records under each of the concepts. For example, an easy way to check whether the singular 'sulfonamide' and the plural 'sulfonamides' fall within the same concept is to enter **Explore References: Research Topic** 'sulfonamide or sulfonamides'. In this case the separate concepts are found to have identical numbers of hits so the singular and the plural are included in a single concept. (For a related case, see Chapter 6, Section 6.7.)

### 3.3.3   Synonyms

SciFinder finds synonyms for terms entered by using algorithms that:

- Convert acronyms to full text (e.g. HIV to human immunodeficiency virus);
- Identify chemical substances and if names of substances are identified then the CAS Registry Numbers are automatically included in the search;
- Refer to a manually built synonym dictionary;
- Identify Index Headings and, if identified, the Older/Newer/Used For terms related to the Index Heading may be included in the search;
- Identify English/American spellings (e.g. **Explore References: Research Topic** on flavor or flavour gives identical numbers of hits);
- Include standard abbreviations when the full term is entered (e.g. **Explore References: Research Topic** 'chromatography' also searches the abbreviation chromatog used in CAPLUS).

However, what constitutes a 'synonym' may depend critically on the context, so SciFinder will not always search terms exactly as intended. The synonym dictionaries and algorithms, which are constantly being reviewed in order to achieve an optimal balance between search comprehension and precision, are not available publicly so the only way to identify terms that have been searched automatically is to look for hit terms in answers retrieved.

Automatic searching of predetermined synonyms greatly assists the user, who nevertheless should still *consider* adding additional synonyms. This is achieved at the **Explore** level either by using the conjunction 'or' or by entering terms in parentheses. For example, **Explore References: Research Topic** 'HIV in humans (men, women)' and 'HIV in humans or men or women' give similar lists of candidates although the list from the latter query gives some additional 'closely associated' options. The differences between the candidates are generally of little consequence since both contain the important candidates where 'hiv' and one of 'humans/men/women' are 'closely associated' and 'anywhere in the reference'.

Synonyms entered under **Explore References: Research Topic** are treated as separate concepts, so if many synonyms are added the list of candidates may be extensive. Searching for information generally requires compromises particularly relating to comprehension versus precision; it helps if the searcher is alert to the various issues and then applies the most appropriate search to meet the information needs.

### 3.3.4   Phrases

Words not separated by prepositions (or conjunctions or stop-words) are identified as a single concept. Answers are retrieved where these words are 'closely associated', and not just as the exact phrase. This is important since searching for exact phrases fails to retrieve answers where the order of the words is different or where other words are in between. Since these factors occur very frequently, searches for phrases will almost invariably miss important answers. Accordingly it makes no difference if 'traditional

Chinese medicines' or 'Chinese traditional medicines' is entered under **Explore References: Research Topic** since the 'concept' retrieves records that contain the three terms at least somewhere in the same sentence (irrespective of the order).

However, while SciFinder thus correctly searches for words within the sentence and not as a phrase, there may be instances where precise phrases are required. To assist the searcher in these cases SciFinder may start the list of candidates with an indication of the number of records 'as entered'. For example, **Explore References: Research Topic** 'traditional chinese medicines' gives two candidates:

> 2072 references were found containing **'traditional chinese medicines'** as entered;
>
> 24,219 references were found containing the concept **'traditional chinese medicines'**.

Answers in the former set contain exactly the words 'traditional', 'chinese', 'medicines' in that order. In the latter set, the answers will additionally include words within a sentence ('closely associated') and words where the SciFinder algorithm has been applied (e.g. 'medicine' as well as 'medicines').[1]

At times records retrieved under candidates 'as entered' will not *exactly* match the entry. For example, **Explore References: Research Topic** 'oil in water' will retrieve 'oil with water' and 'oil water'. The reason is that SciFinder may allow up to one intervening word between the main terms when a preposition is entered in the query. However, this is a detail that experienced users of SciFinder will observe and will understand as an outcome of the algorithm that has been implemented. The developers of these algorithms have in-depth knowledge of the databases and have included in the algorithm the functions that they consider will provide the best answer sets based on the queries entered.

One use of candidates 'as entered' is to allow inclusion of words that SciFinder does not include in concepts. For example, the list of candidates from **Explore References: Research Topic** 'off flavours in wines' gives candidates only with the concepts 'flavours' and 'wines'. While references with these two concepts may be obtained and then examined in turn or narrowed by strategies mentioned later in this chapter, another option is to enter **Explore References: Research Topic** 'off flavours'. The candidate 'as entered' will include exactly 'off flavours' (ca. 100 records) and the references may be chosen and later narrowed to records relating to wines.

However, the candidate 'as entered' does not employ the SciFinder algorithm and so, among other things, American/English spellings, truncation, and singulars/plurals are *not* searched. Since the databases have almost exclusively American English spellings it is better to enter **Explore References: Research Topic** 'off flavors' (ca. 1800 records). It always helps if users employ 'scientific method' throughout the process, i.e. that critical analysis of the results of the initial 'experiment' is conducted (and in this case the user would immediately think why relatively few answers are retrieved for 'off flavours' 'as entered').

### 3.3.5   CAS Registry Numbers

The importance of searching for CAS Registry Numbers for substances was illustrated in Chapter 1, so it is necessary that one of the first actions of the algorithm

---

[1] By comparison, the numbers of answers 'as entered' for the searches 'traditional Chinese medicine', 'Chinese traditional medicine', and 'Chinese traditional medicines' are 12,906, 1342, and 340 respectively. The number of answers for the 'concept' in each case is the same (24,219).)

behind **Explore References: Research Topic** is to determine whether any of the terms entered corresponds to the exact name of a substance in REGISTRY. If this is the case, SciFinder automatically searches the CAS Registry Number as one of the terms within the concept.

However, the key issue is that a name in the substance database is recognized, which is most likely to occur for substances with common or simple names. Unless the user has an excellent knowledge of nomenclature, in more complicated cases the name entered in the query may not match exactly and the CAS Registry Number is not searched.

---

*SciFinder Tip*

If **Explore References: Research Topic** includes a term for a substance, then the search should include its CAS Registry Number. Check a full record to verify that this has occurred.

The other option is to include the name of the substance and its CAS Registry Number in the query. If there are other concepts in the **Explore References: Research Topic** then the list of candidates will include the concepts 'closely associated' with the CAS Registry Number – and that gives an additional level of precision to consider.

---

## 3.4 Choosing Candidates

Once the list of candidates has been obtained, the next task is to choose options from the list. Generally some of the first listed candidates contain more precise answers and have fewer references since they include the greater number of concepts. However, if there are very few records with all the concepts, then either the question needs to be revised or else other candidates with fewer concepts need to be chosen.

On the other hand, if there are a large number of references with all the concepts, then either more concepts must be added to the question or else **Get References** for appropriate candidates should be obtained and further refinements undertaken at a later stage. Since SciFinder has many options for subsequent refinements (Section 3.5), the latter option generally is better and indeed users should not be deterred by initial answer sets of several thousands of records.

Candidates where the concepts are 'closely associated' generally produce more precise answer sets and are often chosen where many references are obtained or when a quick initial review of indexing is required. For example, it may not be productive to work through large answer sets and then finally to realize that some very important Index Headings were not retrieved in the initial **Explore References: Research Topic**.

---

*SciFinder Tip*

In general, it is better to choose candidates based primarily on the search requirements rather than on numbers of references. This is particularly the case since large answer sets may be quickly narrowed with SciFinder post-processing tools.

It is reasonable to choose a smaller answer set in the first instance to quickly check hit terms and to check indexing. However, once this review has been completed it is often advisable to change the initial query or to go back to a larger answer set in order to obtain more comprehensive and more precise answers.

Such processes are often followed in science: initial experiment, review, revise experiment. They should be considered also in information retrieval.

## 3.5   Working from the Reference Screen

The initial reference screen (Figure 3.3) contains many functions, some of which are summarized in Table 3.2. Note that functions may be performed on the entire answer set or on specific answers. For example, clicking **Get Substances** at the References level (marked ❸ in Figure 3.3) retrieves records for all the substances indexed in all the records, while clicking **Substances** at the answer level (marked ❽) retrieves records for substances in the specific record. The same applies for retrieving reaction and citation information at the different levels. The implementation of this latter function is described in Section 6.6.

### 3.5.1   Keep Me Posted ❶

Current awareness searches are set up in SciFinder through the **Create Keep Me Posted** function. When this option is clicked, the screen (Figure 3.4) is displayed; then entries

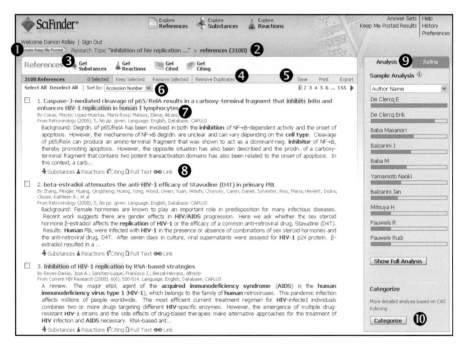

***Figure 3.3***   *Initial reference answer set in SciFinder. By default, references from CAPLUS are presented first, followed by references from MEDLINE. The numbers inserted on this figure are referred to in Table 3.2. SciFinder*® *screens are reproduced with permission of Chemical Abstracts Service (CAS), a division of the American Chemical Society*

**Table 3.2**    *Brief description of functions displayed in Figure 3.3*

| Locator in Figure 3.3 | Function | Brief description | From more details see Section |
|---|---|---|---|
| ❶ | **Create Keep Me Posted** | Sets up a current awareness search. | 3.5.1 |
| ❷ | Search **History** | Tracks session progress; provides quick option to go back several steps. | 3.5.2 |
| ❸ | **Get Substances** **Get Reactions** **Get Cited** **Get Citing** | Retrieves substances, reactions and citations for the entire answer set. | 6.6 |
| ❹ | **Keep Selected** **Remove Selected** **Remove Duplicates** | Keeps/removes selected items (i.e. those with the boxes checked). Removes duplicates from the entire answer set. | 3.5.3 |
| ❺ | **Save** **Print** **Export** | Saves answers in SciFinder. Prints answers. Saves answers on user's computer (different **Save** formats are available). | 3.5.3 |
| ❻ | **Sort** Options | Sorts by Accession Number, Author Name, Publication Year, or Title. | 3.5.3 |
| ❼ | **Link** to Reference Detail | Goes to full database record. | 3.5.4 |
| ❽ | **Links** from Specific Record | Retrieves substances, reactions, citations, and full text for the specific answer. | 3.6 |
| ❾ | **Analysis Refine** | Provides options for refining searches by bibliographic, technical, and index terms. | 3.5.5–3.5.7 |
| ❿ | **Categorize** | Provides advanced options for refining searches by index terms. | 3.5.8 |

are made and **Create** is clicked. SciFinder updates the search every week and sends results to the user.

The current awareness profile does not have to be the same as the final search conducted in a full search of the file. Often in a full file search the user first obtains a large answer set and subsequent refinements are driven by the need to narrow answers to an acceptable number. For example, an answer set of 10,000 answers, obtained through a search over a 25 year period in the database, is unmanageable and refinements are needed. However, this result means that on average one answer is added each day; thus over the period of a **Keep Me Posted** update around seven new hits will be obtained. This is quite manageable. It is therefore suggested that users consider setting up current awareness profiles at a more general level.

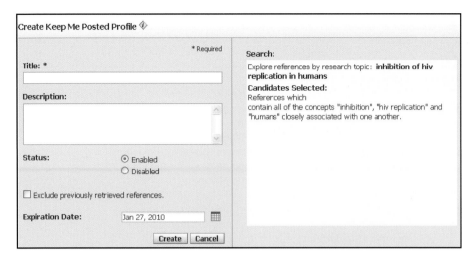

**Figure 3.4   Create Keep Me Posted** *set-up screen. When clicked, KMP inserts the current search into the Search box, but almost invariably broader searches should be considered. SciFinder*® *screens are reproduced with permission of Chemical Abstracts Service (CAS), a division of the American Chemical Society*

### 3.5.2   Search History ❷

As the session continues the search history line next to the **Create Keep Me Posted** link is updated. Figure 3.5 shows changes in the history after the search in Figure 3.3 has been narrowed to records for patents.

The benefit of this to the user is that the progress of the session is apparent (the session history may also be followed by clicking the **History** link at the top right-hand corner of the screen) and it is easy to quickly go back to earlier screens by clicking the links.

### 3.5.3   Selecting, Saving, Printing, Exporting, and Sorting Records ❹ ❺ ❻

Individual records may be selected by checking the boxes to the left of the title. Selected records may be kept or removed by clicking the **Keep Selected** or **Remove Selected** links respectively.

Clicking the links **Save, Print**, or **Export** take the user to separate secondary screens which are self-explanatory. The **Save** option saves the session on the CAS server and saved files may be reopened through SciFinder (saved Answer Sets are accessed through links on the top right of SciFinder screens). The benefit is that the SciFinder session is reopened and it may be further processed through the SciFinder tools. The **Print** option offers Summary and Detail, the difference being that the former gives only the bibliographic information while the latter gives the full record. Finally, answers may be exported to a variety of file formats through the **Export** option. Through working with other Internet search interfaces, users will be familiar with how these functions operate.

**Remove Duplicates** removes MEDLINE records that also appear in CAPLUS. It is best if this is performed at final stages in the search, since the indexing in each of

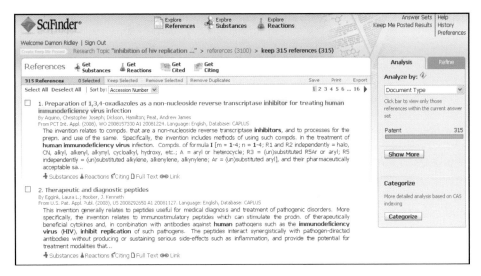

***Figure 3.5*** *Updated breadcrumb. Note the updated analysis on the right (Patents – 315 have been chosen) and the updated breadcrumb (to the right of link to **Create Keep Me Posted**). **Research Topic** Status Line allows users to navigate through the search session quickly. SciFinder® screens are reproduced with permission of Chemical Abstracts Service (CAS), a division of the American Chemical Society*

the databases is different and any post-processing is best done on as much data as possible; when there are duplicate records then refinements may pick up hits from only one of the database records. Of course what constitutes a duplicate depends on the requirement since all records in CAPLUS differ from those in MEDLINE. However, the usual requirement is to remove records that refer to the same original publication, e.g. to obtain a single list of references for which full text needs to be obtained.

The default display order for **Sort** is by Accession Number (the most recently added record first) with CAPLUS records before MEDLINE records. Other **Sort** options are by Author Name (first named author in the document only), Publication Year, or Title, and all of these will intersperse CAPLUS and MEDLINE records. More immediate comparisons may then be made between records in the different databases.

### 3.5.4   Link to Full Record ❼ and Link to Full Text ❽

The title of the article is linked to the full database record (**Reference Detail**). Examples of full records are shown in Figure 1.5 (CAPLUS) and Figure 1.7 (MEDLINE).

**Full Text** links to an electronic version of the full text article. Publishers now provide access to the electronic versions of their journals published from the mid 1990s (but also to backfiles of various lengths) and SciFinder provides direct links to those publishers who have agreements with CAS. If the record is a patent that is available electronically, then SciFinder automatically links to the full text patent.

Access to in-house electronic library collections, to the CAS Document Detective Service[SM] (see Appendix 1 for links), and to other full text options is possible. The process is to click the appropriate check-box at the left of the record and then to click the **Full Text** icon at the top of the screen. However, access to these options is controlled by the user's institution and the library coordinator should be contacted for further details.

The ability to link directly to full text records is a great bonus to users, who thus increasingly have immediate access to the world's scientific literature. As SciFinder links directly with more publishers, who as time passes will have increased numbers of articles available in electronic form, and as libraries obtain greater access to e-journals, the benefits of the integration between the primary and secondary sources will increase.

---

*So Does the Availability of Electronic Full Text Documents Lessen the Need for SciFinder?*

Points to consider include:

- Full text documents need to be obtained from multiple sources, while SciFinder has titles and abstracts for over '36 million' documents in the one location;
- The systematic indexing in CAPLUS and in MEDLINE often enable easy retrieval of important documents that would have been very difficult to search/retrieve through searches of full text – which has issues such as variation in author terminology, length of documents, how to connect terms, and how to search for important new science in the main discussion (perhaps as distinct from searching for words in Introduction and Experimental Sections);
- The REGISTRY database offers multiple ways to search for substances, to find their properties, and to find references (many search options such as structure searching and searching for specific property information cannot be done effectively in full text);
- The CASREACT database allows precise searching of reaction information, which often is represented in nonsearchable graphic form in full text;
- The value of SciFinder post-processing tools for references, substances, and reactions (the first of these is discussed now; the latter two are discussed in subsequent chapters).

SciFinder compliments and adds value to the institution's full text resources!

---

### 3.5.5  Analyze References 

The 12 options under **Analysis** are listed in Table 3.3, and further information is available in the SciFinder help files (accessed by clicking Help at the top right or by clicking an 'i' icon). Seven of these are based on the bibliographic parts of records; the remaining five are based on indexing and are particularly useful for working through *technical aspects* of records.

The structure of **Analyze** is similar in all the options, specifically:

- **Sample analysis** (Figure 3.6(a) shows Sample analysis for CA Section Title) appears in the first screen and provides a list of the most frequent terms. At the bottom of this list there is a link to ...;

**Table 3.3** **Analyze** *options for Reference answer sets in SciFinder. There are seven options based on bibliographic terms and five options based on technical terms*

| **Analyze** option | The histogram shows the number of records: |
| --- | --- |
| *Document-based (bibliographic)* | |
| Database | In CAPLUS and in MEDLINE |
| Author Name (default) | For each author entry, although note a single author may have more than one entry (e.g. last name, followed by initials or by full given names) |
| Company | For each company entry, although note a single company may have many different entries (it depends mainly on what was reported in the original document) |
| Document Type | For each document type (e.g. Journal, Patent) |
| Journal Name | For each Journal Title |
| Language | For the languages of the original documents |
| Publication Year | For the different years of publication |
| | |
| *Index-based (technical)* | |
| CAS Registry Number | For the CAS Registry Numbers indexed |
| CA Section Title | For the various broad technology sections in CAPLUS |
| Index Term | For the Index Headings indexed |
| CA Concept Heading | For the Subject Index Headings |
| Supplementary Term | For single words in supplementary terms |

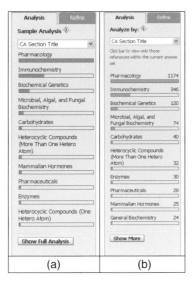

**Figure 3.6** *The screens obtained through **Analyze** options: (a) Initial screen once **Analyze: CA Section Title** is chosen; (b) screen after **Show Full Analysis** is chosen. SciFinder® screens are reproduced with permission of Chemical Abstracts Service (CAS), a division of the American Chemical Society*

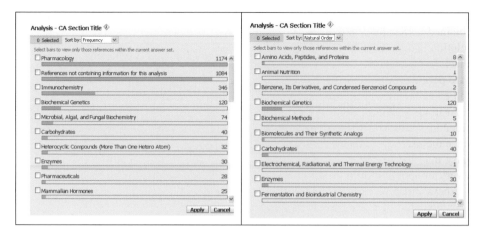

***Figure 3.7*** *Screens after **Show More** (Figure 3.6(b)) is chosen. The default is to show the histogram by Frequency (left), but the list may also be sorted by Natural Order (right), which may allow specific headings to be identified more rapidly. SciFinder*® *screens are reproduced with permission of Chemical Abstracts Service (CAS), a division of the American Chemical Society*

- **Show Full Analysis** (Figure 3.6(b)), which additionally shows the number of records for the most frequent terms. At the bottom of this list there is a link to ...;
- **Show More** (Figure 3.7), which now shows the first part of the list through which the user may scroll. Boxes appear on the left of these terms and after they are checked, the user clicks **Apply** and the answer set is narrowed to those which contain the marked terms.

**Analyze** searches through answers and provides the user with information that can be studied carefully before action needs to be taken. The ability to analyse actual information in records greatly helps to overcome issues relating to the variability of terms in free-text written by authors, and reduces the need for prior knowledge of indexing.

**Analyze** options have different uses and at times there are some factors that need to be considered. Since Figures 3.6 and 3.7 show **Analyze: CA Section Title**, the uses and additional considerations for this option are discussed first.

### 3.5.5.1   CA Section Title

The CAPLUS database is divided into 80 broad sections, where a section is shown in each full record (e.g. the record in Figure 1.5 comes from Section 10-2: Microbial, Algal, and Fungal Biochemistry). A summary of the current sections is available in Appendix 1 and more detailed information is available through links from this website.

However, CA Sections have changed over the 100 years covered in CAPLUS so the list obtained may contain more than 80 terms. Changes in indexing always need to be considered; the user may find one section of particular interest, but when answers are obtained it may be found that they are only from an earlier period of the database! The solution is to look through the entire list, to select all those of potential relevance, and to check some of the first and last answers for publication years (to verify that the required time period has been covered). The new answer set then may be analysed in other ways and eventually the most appropriate set of answers is obtained.

At times it may be preferable to **Sort** the analysis list by Natural Order (alphabetical order in the case of text entries), since it may be easier to detect relevant entries and since the user is less focused on numbers of records. A recently added and precise index entry may have relatively few hits, but because of its currency and precision it may be a very important term to include.

Another consideration with CA Section Titles is that, while a single section code is entered for each record, the science in the original document may also have been represented by another section code. Again the solution is to consider each entry in the list carefully and to narrow answers in stages.

### 3.5.5.2   *Database*

At some stage the user needs to decide whether to continue the analysis on the full bibliographic answer set or in the individual databases. Particularly because the indexing in the databases is different, it may at first be confusing to see CAPLUS indexing interwoven with MEDLINE indexing. Further, CAPLUS and MEDLINE may have different policies relating to entry of bibliographic information and a particular example is that historically CAPLUS contains the names of authors actually entered in the original document (full names or last name and initials) whereas until very recently MEDLINE contains the last name and initials only (even when full given names were in the original document).

In the cases of **Analyze: CA Section Title**, of **Analyze: CA Concept Heading**, and of **Analyze: Supplementary Term**, all analysis is done in CAPLUS, so there is no complication. However, in other cases it may be better to continue the analysis in the individual databases and to combine the answers at a later stage (and finally remove duplicates). **Analyze: Database**, and then the choice of each database in turn, gives records in the individual databases.

### 3.5.5.3   *Other Analysis Options*

Once it is understood that **Analyze** previews entries under a number of options and enables choices to be made based on the actual information in the database(s), and it is understood that there is the need to work through the **Analyze** secondary screens **Show Full Analysis** and **Show More**, then it suffices at this stage to comment on prospective uses of **Analyze** and to note some general issues. However, users will discover other issues from time to time and future searches should allow for them.

Table 3.4 summarizes document-based (bibliographic) options, while Table 3.5 summarizes index-based options, which are particularly useful for searches on scientific topics.

**Table 3.4**    *Summary of document-based (bibliographic)* **Analyze** *options*

| Analyze option | Uses | Notes |
|---|---|---|
| Database | To identify numbers of records in each of CAPLUS and MEDLINE. | At times one of the first actions may be to obtain answers for the databases separately (e.g. to explore database specific index terms by **Analyze: Index Term** or **Categorize**). |
| Author Name | To narrow records to those from particular authors.<br>To see variations in entries for individual authors (there can be surprises!) | Author names entered in CAPLUS are those in the original document.<br>Records in MEDLINE prior to 2006 contain last name and initials only.<br>There may be several variations to consider. **Sort: Natural Order** gives an alphabetical list and allows single authors to be recognized more readily. |
| Company Name | To narrow records to those from particular organizations.<br><br>To survey what competitor organizations are reporting. | Company names may change at times (e.g. through acquisitions and mergers).<br>There is considerable variation in the way a single company may be listed. It depends on what name the authors used and whether authors included their department name, or only the main organization, or ZIP codes, or whether some or all of the entries were abbreviated or used acronyms (e.g. NASA). |
| Document Type | To understand the different types of documents in the answer set (e.g. journals, patents, reviews, conference records) and to choose the ones of particular interest. | Records for patents occur in CAPLUS, but not in MEDLINE; MEDLINE has a number of additional document type tags such as Clinical Trials. |
| Journal Name | To limit answers to certain journals (e.g. journals considered of most interest in the field or journals for which electronic or print versions may be most readily obtained).<br>To find the important journals in the field (e.g. to recommend their acquisition or to consider them as a possibility for submission of a document for publication). | There are some differences in the listing of journals in CAPLUS and in MEDLINE.<br>Journal Titles may have changed over time, some may be discontinued, and some may have started publication only recently – so to narrow answers by Journal Name may also restrict answers to certain time periods. |

**Table 3.4** (*continued*)

| Analyze option | Uses | Notes |
|---|---|---|
| Language | To limit answers to language of original publication, and so to exclude other languages. | Patent families in CAPLUS appear under a single record and the language stated in the record will be that of the initial patent family member reviewed by CAS. While the original patent may not be in English, members of the Patent Family in English may be available; individual records need to be checked. |
| Publication Year | To limit answers to publication years. | This option is often chosen to narrow answers to a manageable size. More recent research may be more relevant in some technologies. |

**Table 3.5**   *Summary of index-based (technical)* **Analyze** *options*

| Analyze option | Uses | Notes |
|---|---|---|
| CAS Registry Number | To find most listed substances in records (e.g. to understand the most significant substances mentioned on the topic). To find CAS Registry Numbers (e.g. when **Substance Identifier** does not retrieve required substance). | See Chapter 4 for further information on identification of CAS Registry Numbers. |
| CA Section Title | To narrow records to broad technology areas. | CA Section Titles may change with time. |
| Index Term | To find Index Terms in all the records. | Default listing is by frequency; alternative listing is in alphabetical order. Index Terms may change with time, so Index Terms relevant to the time period of interest must be chosen. |
| CA Concept Heading | To limit answers by CA Subject Headings. | Subject Headings may change with time. Subject Headings provide precise and comprehensive entries to technical topics. |
| Supplementary Term | To limit answers to author-related technical terms. | Supplementary terms are entered by CAS document analysts to reflect author terms. Most Supplementary Terms are single words only. |

---

*SciFinder Tip*

**Analyze** allows options to be considered before a choice needs to be made. Therefore, start with general searches, analyse outcomes, and make refinements subsequently.

Use bibliographic-based **Analyze** options when bibliographic outcomes (e.g. Company Name or Publication Year) are most important. Use index-based **Analyze** options to narrow answers by systematic technical terms. The variety of technical **Analyze** options in SciFinder puts SciFinder well ahead of all other scientific search interfaces!

Try different options. Evaluate results carefully. **Analyze** is a very important function and should nearly always be explored as an option to narrow answers from initial searches.

---

### 3.5.6   Refine References

The seven options under **Refine** and the types of entries that need to be made are listed in Table 3.6, and further information is available in the SciFinder help files (e.g. through links given in Appendix 1). Six of these options are based on the bibliographic parts of records; the remaining one, **Refine: Research Topic**, searches the query in titles, abstracts, and indexing, and uses the SciFinder **Explore References: Research Topic** algorithms.

The document-based **Refine** options are best applied when the user is sure of the type of information that is required and hence does not need to view **Analyze** options first. Note also that many of these options may be chosen right at the start of **Explore References: Research Topic** through the additional options shown in Figure 1.1. These functions are available in other Internet search engines and there is a tendency for users to choose them at the beginning, but the extensive SciFinder tools available usually offer better ways to narrow answers and hence to apply these at later stages.

**Refine** provides one technical option not available under **Analyze**, specifically **Refine: Research Topic**. Here terms in the title, abstract, and index fields are searched, whereas the technical **Analyze** tools (Table 3.5) search entries in the index fields only.

It may appear that an alternative way to search terms in titles and abstracts is to go back to the original **Explore** statement, but this may produce quite a complicated list of candidates – none of which really meets the key search need. For example, consider a search in which there are four concepts A, B, C, D, where A always needs to be closely associated with B, where C is a synonym for 'A with B', and where D separately needs to be anywhere in the record. Perhaps the search can be stated [(A with B) or C] and D. An easy way to sort this out is to do an initial **Explore References: Research Topic** 'A with B or C' and then a subsequent **Refine: Research Topic** 'D'.

To illustrate such a case, one important property of polymers is the ratio of molecular weight to molecular number (Mw:Mn), but information on this very specific technical issue is likely to occur mainly in the abstract text. Accordingly, if information on this ratio for linear low density polyethylenes (lldpes) is required, **Explore References: Research Topic** 'linear low density with polyethylene or lldpe' is first conducted and the initial relevant answers (>16,000) are then narrowed with **Refine: Research Topic**

**Table 3.6**    **Refine** *options for bibliographic records in SciFinder. When the* **Refine** *option is clicked, boxes that request additional information appear. This table gives a summary of the types of entries required*

| **Refine** option | Example of entry required |
| --- | --- |
| *Document-based (bibliographic)* | |
| Database | Choices are:<br>- CAPLUS<br>- MEDLINE |
| Author Name | Entry of Author Last Name is required.<br>Entry of First/Middle Names or initials are optional. |
| Company Name | Entry of full name or part name of Company is required. |
| Document Type | Options are:<br>- Biography          - Historical<br>- Book                 - Journal<br>- Clinical Trial      - Letter<br>- Commentary      - Patent<br>- Conference        - Preprint<br>- Dissertation       - Report<br>- Editorial            - Review |
| Language | Options are:<br>- Chinese            - Japanese<br>- English             - Polish<br>- French              - Russian<br>- German             - Spanish<br>- Italian |
| Publication Year | Entry of Publication Year(s) is required. |
| *Technical* | |
| Research Topic (default) | Research Topic query box appears and text entries are made. |

'mw mn'. This gives around 350 answers, nearly all of which contain actual data for the Mw:Mn ratio.

> *SciFinder Tip*
>
> There may be cases where it is preferable to search only some of the concepts in the initial **Explore References: Research Topic** and to enter additional concepts at **Refine: Research Topic** later. The key is to understand the difference between author and index terms and to search for terms in the most appropriate combinations.

Like all other **Refine** and **Analyze** options, **Refine: Research Topic** operates at the 'and' (or 'not') level and it is not possible to **Refine: Research Topic** in such a way that the terms now entered under refine are 'closely associated' with terms previously searched. The implication is that some level of precision may be lost. Thus if two concepts A, B are entered at the **Explore** level and the candidate in which they are 'closely associated' is chosen, and if the answers are refined with a third concept C, then the outcome would effectively be an answer set (A closely associated with B) AND C. On the other hand, entry of three concepts A, B, C at the Explore level would give, among other things, a candidate where all three are closely associated.

Actually the inability to associate terms closely under **Refine: Research Topic** is a restriction that applies to this option only. All other **Refine** and **Analyze** options necessarily have to be performed at the 'and' level since they involve separate fields for the data being searched.

### 3.5.7  Analyze or Refine?

Several of the options under **Analyze** are also available under **Refine** and the question is which one to use and when.

When a definite requirement is known at the outset then **Refine** (or choice of the Advanced Options in the initial search) gets this result more rapidly. Therefore, if the user knows that the most appropriate next step is to go to records only in CAPLUS, then **Refine: Database, CAPLUS** may be the better option. The same may apply when a certain document type (e.g. patent) or original document language (e.g. English) is required. In other cases it is usually better to **Analyze** first, in which case the various entries may be examined and a more informed choice may be made.

This particularly applies to options like **Analyze: Author Name** and **Analyze: Company Name**, where a single author or company name may be listed in a number of different ways. For example, as seen in Figure 3.3, one of the principal authors in the field of HIV inhibition is Erik de Clercq. **Analyze: Author Name** shows that he is variously listed as De Clercq Erik, De Clercq E, Clercq E De, Clercq Erik De, de Klerk E, and Declercq E. These variations are apparent under **Analyze**, and the relevant boxes may be checked; knowing about, and allowing for, such variations under **Refine: Author Name** or **Explore: Author Name** (Section 6.2) is quite a different matter.

This also applies to **Publication Year**, where the user may think 'I just want information from 2005 onwards' since this is probably driven by the perception that 'just the last few years' will give a manageable number of answers. However, consider **Analyze: Publication Year** first and check all the options; it may be that including a few earlier years still gives a manageable number of answers and perhaps the most relevant answer may have been in 2004.

**Refine** executes instructions exactly as the user specifies, while **Analyze** gives the user a histogram, which indicates outcomes for the various paths that may be followed. Too often users jump to conclusions about what to do next, but remember all those subtle differences in records (Figures 1.5 and 1.7) for the same original document. Consider further the even greater variation in records for different documents! Accordingly, experienced users accept the fact that *they do not know exactly* what is in the databases, and so they almost invariably take advantage of any options that guide them to alternatives. Therefore in most cases experienced users prefer **Analyze**.

While the main reason for choosing **Analyze** is because the user is guided through alternatives, another reason is that **Analyze** allows combinations of alternatives to be chosen. For example, consider the choices the user has if restriction of the records to patents *and* journals only is required. In fact this may be achieved most easily in a single new answer set through **Analyze**. Thus when **Analyze: Document Type** is chosen, a histogram appears and both Journals and Patents may be selected. Indeed, whenever options under **Analyze** are chosen histograms appear from which one or more options may be selected, with the added advantage of knowing the outcomes in advance.

### 3.5.8    Categorize ❿

While **Analyze:  Index Term** is an extremely valuable tool, there are some potential limitations. First, the whole concept of **Analyze** is to *narrow large answer sets*, but large answer sets are likely to yield a large number of index entries, which in turn may take some time to work through. Second, the most frequently listed index entries may be very general (e.g. Index Heading 'Humans' may be posted in many of the records in medical-related studies) so it is necessary to go further down the list. Third, in some cases the histogram obtained may not be immediately informative (e.g. **Analyze:  CAS Registry Number** produces a whole list of CAS Registry Numbers and since those numbers alone do not have any scientific meaning links to the different substances need to be made in turn).

The solution is to use **Categorize**, which operates in a few steps specifically that SciFinder:

- Finds all the Index Headings;
- Sorts them into pre-defined category headings;
- Sorts entries under the category headings further into categories;
- Displays the index terms in the categories.

The twelve pre-defined Category Headings, and the Categories within each of them, are listed in Table 3.7. Further information is available initially through links given in Appendix 1 and then through links given below.

SciFinder uses a number of algorithms to assign the various levels of categories and subject index headings; for substances the categories are assigned in part on the basis of CAS Roles (Chapter 2, Section 2.1.2). There is little need to know how the algorithms operate, and indeed it is more important to study the information displayed in the various categories, to think carefully about the implications, and to try a few different options.

Table 3.8 shows some outcomes when **Categorize** is chosen for results from **Explore References:  Research Topic** 'inhibition of hiv replication in humans' and when initial results are refined to those published since 2002. The first column shows **Categorize** for CAPLUS answers, while **Categorize** for MEDLINE answers are shown in the second column. There is a lot of information in this table, but it gives a quick feel of the types of outcomes and from this it is evident that a few different options should be explored.

A considerable amount of systematic (indexed) information may quickly be viewed and then used in a number of ways. First, it may be useful to check why answers were retrieved. That is part of the scientific way of thinking since at times understanding what happened in experiments and hence being guided to do other experiments may be more important than accepting the result itself. For example, the information line in the last row from MEDLINE in Table 3.8 lists Genetics & protein chemistry > Proteins & peptides >3 selected, and it is worth checking a result. Figure 3.8 shows one of these records. HIV replication, inhibits, and human are in the last sentence in the abstract (which meets the 'closely associated' requirement, and HIV integrase inhibitors (which was one of the Index Terms chosen) is confirmed as an MeSH heading.

**Table 3.7**  *Category Headings and Categories in SciFinder. For example, under the Category Heading: Biology, there are eight defined categories and within each of these are many related index headings*

| Category Heading | | Categories |
| --- | --- | --- |
| All | Substances | Topics |
| | Searched substances | |
| Analytical chemistry | Analysis | Analytes and matrixes |
| | Reagents and other substances | |
| Biology | Anatomy | Animal pathology |
| | Endocrinology | Immunology |
| | Organisms | Processes and systems |
| | Substances in adverse effects | Substances in biology |
| Biotechnology | Agriculture | Food |
| | Medicine | Substances in adverse effects |
| | Substances in agriculture | Substances in biological uses |
| | Substances in food chemistry | Substances in medicine |
| | Toxicology and forensics | |
| Catalysis | Catalysis | Catalysts |
| Environmental chemistry | Astronomy | Environment |
| | Formed, removed, and other substances | Pollutants |
| | Geology and soil chemistry | |
| | Substances in geology and astronomy | |
| General chemistry | General science topics | Inorganic substances |
| | Miscellaneous substances | Organic substances |
| Genetics and protein chemistry | Genetics | Miscellaneous substances |
| | Nucleic acids | Protein and peptide topics |
| | Proteins and peptides | |
| Physical chemistry | Atomic and molecular phenomena | Electric and magnetic phenomena |
| | Gas, liquid, and solid phenomena | Mechanics |
| | Miscellaneous substances | Particle phenomena |
| | Quantum mechanics | Spectra and spectroscopy |
| | Subatomics | Substances in processes |
| | Substances in property studies | Surface phenomena |
| | Thermodynamics | |
| Polymer chemistry | Applications and phenomena | Miscellaneous substances |
| | Modifiers and additives | Polymers |
| | Processes and apparatus | |
| Synthetic chemistry | Bio-prepared substances | Combinatorial reactants and other substances |
| | Combinatorially prepared substances | |
| | Manufactured substances | Prepared substances |
| | Purified substances | Reactants and reagents |
| | Reactions | |

**Table 3.7** (*continued*)

| Category Heading | | Categories |
|---|---|---|
| Technology | Ceramic | Construction |
| | Formed, removed, and other substances | Imaging and recording |
| | | Metallurgy |
| | Materials and products | Processes and apparatus |
| | Power and fuel topics | |
| | Substances in technology | |

**Figure 3.8** *Record in MEDLINE after HIV integrase inhibitors has been selected under* **Categorize***. It helps to see actual hit terms to verify that the types of answers required are being retrieved. SciFinder screens are reproduced with permission of Chemical Abstracts Service (CAS), a division of the American Chemical Society*

Second, answers should be checked to determine whether alternative terms should be used in the search. In this case, the Index Heading 'HIV Integrase Inhibitors' appears and a new **Explore References: Research Topic** 'HIV Integrase Inhibitors' may be considered (i.e. the terms 'replication in humans' may be omitted from the original **Explore** and replaced with 'integrase').

Third, users should always look out for new research areas and in the various displays in Table 3.8 many terms appear that may well be worth following. This is just an example; it is important that the user applies the principles in the actual search being conducted.

**Table 3.8** *Examples of outcomes of* **Categorize** *for* **Explore References: Research Topic** *'inhibition of hiv replication in humans'. 1st column: CAPLUS, 2nd column: MEDLINE. Screens are provided here to give users a quick idea of the types of information available. SciFinder® screens are reproduced with permission of Chemical Abstracts Service (CAS), a division of the American Chemical Society*

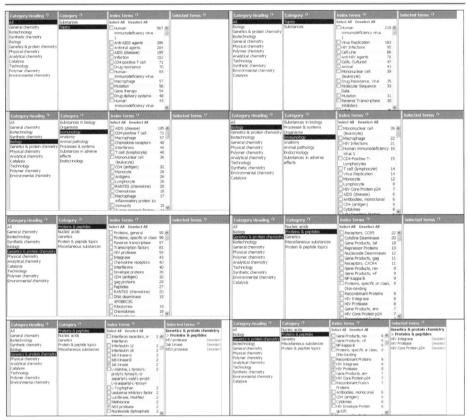

---

**Comment**

When electronic versions of journals first came out, many scientists were reluctant to engage them since the opportunities of 'serendipitous browsing' in print editions were limited. However, the **Analyze** and **Categorize** tools in SciFinder provide opportunities for browsing in a systematic way – and within the broad field of science of interest!

---

## 3.6 Working from the Record Screen

The screen for a *single record* (**Reference Detail**, Figure 3.9) contains links such as **Get Substances, Get Reactions, Get Cited**, and **Get Citing**, which are available also

***Figure 3.9*** *Screen for a single record in CAPLUS (only the first three of 37 citations in the record are shown). Links to **Get Substances, Get Reactions, Get Cited, Get Citing**, and **Get Full Text** apply to the single record only. SciFinder® screens are reproduced with permission of Chemical Abstracts Service (CAS), a division of the American Chemical Society*

through the screen for the references (e.g. Figure 3.5). The difference is that the links from the reference screen apply to all the references and not just to the single record. (The implementation of these functions is described in Section 6.6.)

Clicking the item **Link** (to the left of the **Save/Print/Export** options) provides a url (Figure 3.10), which may be copied and then entered into other documents. If the link is pasted in an email to a colleague, the recipient may click it directly, in which case the user's SciFinder Sign In screen appears, and after sign-in the SciFinder screen for the record appears automatically. The url may also be pasted into in-house databases,

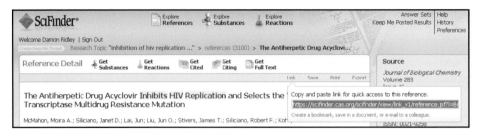

**Figure 3.10** *Display when the item* **Link** *is clicked. The url for the record may be pasted into other documents and the SciFinder record may be directly accessed through this link. SciFinder*® *screens are reproduced with permission of Chemical Abstracts Service (CAS), a division of the American Chemical Society*

but in all cases direct access to the url may be achieved only by colleagues who have access to SciFinder.

There are two different types of links in the Indexing for single records in CAPLUS and they operate differently. When links from the various Subject Index Headings under **Concepts** are chosen, SciFinder searches for all entries in CAPLUS with this Index Heading and displays the records. The advantage is that records for original documents in the same field (i.e. specific Index Heading) are retrieved and these records may then be post-processed.

---

*SciFinder Tip*

A single Index Heading may cover a number of different terms used by authors for the same concept. Not all of these terms may have been included in the initial search, and so the ability to search quickly for all records with a specific Index Heading may give more comprehensive answers. The answers are also precise, since Index Headings are entered in database records only when important new science in the area is reported in the original document.

However, a single Index Heading may appear in several thousand records, so usually it is necessary to narrow answers, which is easily done through SciFinder **Analyze/Refine/Categorize** tools.

---

On the other hand, when a CAS Registry Number below the header **Substances** is clicked, SciFinder displays the single substance record in REGISTRY. Note that the CAPLUS record (Figure 3.9) contains the CAS Registry Number 9068-38-6, which is an example of the many cases where a name is not associated with the CAS Registry Number in the database record. This emphasizes the importance of including CAS Registry Numbers as search terms; in this case it is a systematic term for an enzyme that has 16 different names (Figure 3.11).

If the CAS Registry Numbers under Substances (Figure 3.9) are clicked separately then the individual REGISTRY records are displayed in turn. However, if **Get Substances**

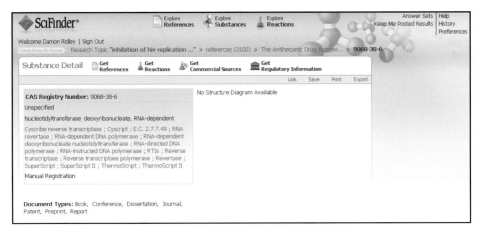

**Figure 3.11** *Substance identifying information (Substance Detail) for the record in REGISTRY, which is obtained after the CAS Registry Number 9068-38-6 in Figure 3.9 is clicked. This REGISTRY record also contains extensive bibliographic and property information (not shown in this figure). SciFinder® screens are reproduced with permission of Chemical Abstracts Service (CAS), a division of the American Chemical Society*

(i.e. the link next to Reference Detail in Figure 3.9) is clicked, a dialog box (Figure 3.12) appears and once options are chosen a summary of *all* the indexed substances is obtained (Figure 3.13).

 If options in Figure 3.12 are chosen then the substances will be narrowed to specific studies, but is it noted that **Analyze: Substance Role** is the default in Figure 3.13. In effect Figure 3.12 offers the **Refine** option while Figure 3.13 gives the **Analyze** option. The issues outlined in Section 3.5.7 (**Analyze** or **Refine?**) for references apply equally here for substances; in most cases it is preferable to look at **Analyze** first and then make informed decisions based on the actual information in the databases. Naturally, the analysis would need to be done on all substances.

## 3.7  Applying Scientific Method to Information Retrieval

This chapter has described the fundamentals of using **Explore References: Research Topic** and has outlined many of the ways that SciFinder works behind the scenes to guide the user. However, it helps if the searcher applies 'scientific method' to information retrieval, and in particular works through a series of steps:

Step 1. Conceptualize the initial search query.
Step 2. Perform an initial search.
Step 3. Carefully examine the initial answers.
Step 4. Revise the search query based on an analysis of initial answers.
Step 5. Explore alternative search options.

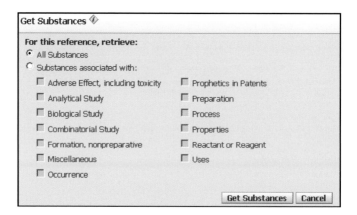

**Figure 3.12**    *When **Get Substances** is chosen, options to retrieve all substances or selected substances are available followed by information relating to CAS Roles in CAPLUS bibliographic records for the substance(s). SciFinder® screens are reproduced with permission of Chemical Abstracts Service (CAS), a division of the American Chemical Society*

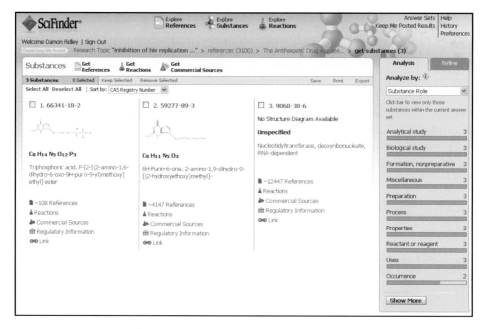

**Figure 3.13**    *Substance answer sets (records in REGISTRY) in SciFinder. SciFinder® screens are reproduced with permission of Chemical Abstracts Service (CAS), a division of the American Chemical Society*

To illustrate these steps, consider a question to find information on 'drugs for the treatment of AIDS'. Some thought processes and search options are discussed below, and actual search inputs and outcomes are summarized.

### 3.7.1   Step 1. Conceptualize the Initial Search Query

Thought processes may include:

- The CAPLUS and MEDLINE databases cover this topic from basic research through to clinical treatments, so SciFinder is an excellent place to start;
- 'AIDS' is an acronym, so the full term acquired immunodeficiency syndrome and other synonyms need to be considered;
- Synonyms for 'treatment' need to be considered;
- 'Drugs' is a generic term, synonyms (e.g. pharmaceuticals) need to be investigated, and ultimately the actual substances involved need to be found;
- SciFinder offers ways to find actual substances from bibliographic answer sets through **Categorize** and through **Get Substances**, so, as search terms for drugs/pharmaceuticals would be hard to determine at the outset, the concepts 'treatment' and 'AIDS' are investigated first.

### 3.7.2   Step 2. Perform an Initial Search

Actions and results:

- Sign in to SciFinder;
- **Explore References: Research Topic** 'treatment of AIDS';
- Research Topic Candidates are shown in Figure 3.14;
- Choose references with treatment of AIDS 'as entered'.

Thought processes:

- Over 37,000 results with the two concepts 'closely associated'. This is a large number, so consider the more precise option 'as entered';

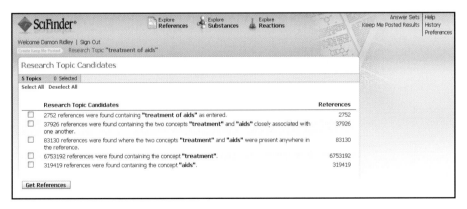

**Figure 3.14**  *Candidates from* **Explore References: Research Topic** *'treatment of AIDS'. SciFinder® screens are reproduced with permission of Chemical Abstracts Service (CAS), a division of the American Chemical Society*

- Recognize that 'as entered' is quite limiting (i.e. definitely not comprehensive). However, the initial aim is to find key Index Headings; an answer set of over 2700 should be sufficient for this purpose.

### 3.7.3   Step 3. Examine the Initial Answers

Thought processes:

- It is possible that the indexing in CAPLUS is quite different from the indexing in MEDLINE, so it is probably a good idea to obtain and explore answers separately in each database.

Actions and results:

- **Analyze** or **Refine** Database:

  - Answers from CAPLUS (1701) and MEDLINE (1048) are indicated;
  - No need to Remove Duplicates now because separate investigations for indexing are planned and Remove Duplicates would reduce MEDLINE answers further;
  - Obtain answers from CAPLUS first.

- **Analyze:   CA Section Title** shows most listed sections are Pharmacology (705 answers) and Pharmaceuticals (159 answers);
- Since the question asks for 'drugs', choose the 159 answers under the CA Section Title: Pharmaceuticals and then **Categorize**;
- Many options may be investigated, but Category Heading Biotechnology and then:

  - Category Medicine indicates most listed Index Terms are Drug Delivery Systems (43), Antiviral agents (40), and Anti-AIDS agents (36);
  - Category Substances in Medicine indicates many 'drugs' (e.g. Zidovudine) and substances for drug formulation.

Thought processes:

- These results were obtained from a very specific initial search, and there is little point in pursuing them further (further investigations should now be conducted on a more comprehensive answer set);
- The important Index Heading in CAPLUS is Anti-AIDS agents and this should be explored;
- However, first check the MEDLINE indexing.

Actions and results:

- Go back to initial answers (2079) and **Refine: Database** MEDLINE, which gives 1048 answers;
- **Analyze: Index Term** indicates headings such as Acquired Immunodeficiency Syndrome, DT (Drug Therapy), and TU (Therapeutic Use);
- A check of an actual record shows the MEDLINE indexing Acquired Immunodeficiency Syndrome DT, drug therapy.

Thought processes:

- Index heading in MEDLINE uses 'therapy' rather than 'treatment', which was used in the original search;
- Revised search in MEDLINE must be undertaken.

### 3.7.4   Step 4. Revise Search

Actions and results:

- **Explore References:Research Topic** 'anti AIDS agents' gives 24,000 references 'as entered' and 60,000 references for the concept;
- **Explore References:Research Topic** 'acquired immunodeficiency syndrome with drug therapy' gives 5600 references for the two concepts 'closely associated'.

Thought processes:

- Important indexing in CAPLUS and in MEDLINE has been considered, but the numbers of references needs to be narrowed – particularly since a maximum of 15,000 may be processed through **Categorize**;
- There is now a need to work through other post-processing options.

This is a big topic, but big problems are best solved in smaller steps. The principles of initial search followed by understanding of the indexing have been followed. The challenge now is to narrow answers in systematic ways.

Searches in SciFinder may indeed be performed in a 'quick and simple' way. However, more comprehensive and precise results may be obtained through an understanding of the database content and of how the search engine works. It may be challenging to decide which path to follow when large answer sets are obtained, but SciFinder post-processing tools offer many options.

## 3.8   Summary of Key Points

- SciFinder is the front end to databases of many millions of records that contain text from authors and indexers. When a search is constructed it helps if the user considers what terms an author and an indexer may have entered. In particular, it should be remembered that authors do not follow any universal policies in writing up text in titles and abstracts, and that there may be very considerable variations in the way different authors write about their work;
- SciFinder guides the searcher through the investigation. The user is first guided by a list of research topic candidates and second through **Analyze** and **Categorize**. In this way SciFinder helps the user overcome many of the issues that arise because of the size and complexity of the databases;
- Central to the operation of SciFinder is the option to **Explore References: Research Topic** using simple natural language questions;
- There are a number of considerations relating to how to enter the initial query, and these are summarized in Appendix 3;

- SciFinder identifies different concepts based on the presence of the prepositions, conjunctions, and stop-words in the query, displays candidates where the concepts are within the same sentence ('closely associated') or simply within the record ('anywhere in the reference'), displays candidates with combinations of the individual concepts, and displays the individual concepts;
- Individual answers should be checked for hit terms and modifications should be made to the query if important Index Headings become apparent;
- **Analyze** and **Categorize** provide many powerful options for revision of answer sets by systematic index terms. **Refine** requires direct entry of terms by the searcher.

# 4

# Explore by Chemical Substance

## 4.1 Introduction

There is hardly any scientific discipline that does not embrace chemicals! Chemicals may be relatively simple like the individual atoms or may be complex like a single DNA molecule that has many billions of atoms. Creating a database for the vast numbers and varieties of chemicals is a major challenge!

However, the issues go further than that. For example, sometimes:

- The bonds that hold atoms together are difficult to describe in computer systems (e.g. resonance, $\pi$-bonds);
- A single substance is referred to by many different names;
- Substances are incompletely described in the literature (e.g. 'xylene' and 'alanine' are not precise descriptors – the former has regioisomers and the latter has stereoisomers);
- Substances may be loosely described (e.g. the agriculturist may refer to N:K:P ratios in soils and is talking about something very different from the actual elements nitrogen, potassium, and phosphorus).

Descriptions of molecular structures occupy large sections of tertiary chemistry courses. Remember the lectures: ionic and covalent bonding, valence bond and molecular orbital theory, resonance and tautomerism, coordination compounds, molecular associations, the structures of metals and alloys, the formation and structures of polymers? Tied in with this are the complexities of nomenclature, and it is apparent that there are many issues for those who build comprehensive substance databases to consider. (Some of these issues are discussed in 'The Challenges with Substance Databases and Structure Search Engines', in *Australian Journal of Chemistry*, 2004, **57**, 387–392.)

Over the years, scientists at CAS have addressed the issues involved with the description of substances and, working with scientists worldwide, continue to address new issues as they arise. For example, combinatorial chemistry and supramolecular chemistry are two relatively recent developments that challenge any indexing system for substances.

*Information Retrieval: SciFinder®, Second Edition* Damon D. Ridley
© 2009 John Wiley & Sons, Ltd

The aim is to produce as simple and as systematic a description of substances as possible, and then to index substances in such a way as to facilitate comprehensive and precise retrieval of information.

This is achieved through entry of substances in a chemical substance database in which each unique substance is given a single registration number. The so-called CAS Registry Numbers are now used widely to identify substances and the master collection of disclosed chemical substance information (REGISTRY) that contains all CAS Registry Numbers is available through SciFinder.

CAS Registry Numbers may be found in REGISTRY in many ways, but most commonly through chemical structure, molecular formula, or chemical name based search terms. These choices help overcome the problems associated with names of substances, and once the CAS Registry Numbers have been identified any search based on them will help overcome the problems associated with the use of many different names for the same substance.

This chapter describes key aspects of the Registry System and then discusses how to find CAS Registry Numbers by searches based on exact structure, name, molecular formula, and keyword terms.

## 4.2    Registration of Substances

Where possible, the CAS Registry System uses the valence bond theory for atoms and so structure representations are mostly the same as those that scientists normally use. However, simple valence bond theory may be inadequate or may be interpreted differently by scientists, in cases such as resonance, tautomerism, $\sigma$-bonding, and $\pi$-bonding. Valency bond theory also has limitations in describing structures for many classes of substances such as polymers, cyclophanes, and other supramolecular assemblies. Thus, sometimes CAS needs to apply special policies for representations of structures.

---

*Comment*

Initially the indexing of some substance records in REGISTRY may be confusing. However, there are several points to note. First, some substances are difficult to describe, particularly in the consistent way required for electronic databases - computers are not very tolerant to 'interpretations' or variability. Second, scientists may not always be rigorous in the way they describe substances (e.g. what is the substance xylene and how should it be indexed?). Third, policies for registrations of substances may need to be revised as the science develops (and this applies in particular to polymers, alloys, ceramics, and the substances in biology).

Always the solution is to consider the 'confusing' registrations carefully; i.e. look through records, try to understand why they appear as they are, and then apply this knowledge in constructing future search strategies.

---

### 4.2.1   CAS Registry Numbers

CAS Registry Numbers are unique descriptors for chemical substances. They are assigned in chronological order and so the numbers have no chemical significance.

A CAS Registry Number is given to *each unique substance*, so a single amino acid variant in two proteins requires two different registrations. Similarly, the sodium and the potassium salts of a carboxylic acid will have different CAS Registry Numbers, and these in turn will be different from the parent acid. However, in some cases different forms or modifications of a substance will have the same CAS Registry Number. For example, the CAS Registry Number for a polymer is generally based on the starting components; the method of polymerization (such as conditions and catalysts) is not considered in the registration.

CAS Registry Numbers are used systematically in CAS databases to index substances (see Table 2.1 for a summary of the policies for entering CAS Registry Numbers in CAPLUS), so searches on specific substances in CAS databases nearly always should involve searches on CAS Registry Numbers. The two instances where other approaches are necessary are:

- For extracts from natural sources (e.g. natural oils and fats; Appendix 4, Section A4.5.3);
- For very recent entries to CAPLUS where indexing is incomplete (i.e. CAS Registry Numbers have not yet been added to the records).

On the other hand, there are occasions when substances that have CAS Registry Numbers will not have any literature citations. There are a number of reasons for this, including:

- Organizations may apply for CAS Registry Number registrations even though they have not published data on the substances;
- Certain parent ring structures have CAS Registry Numbers;
- Newly added substances where the CAPLUS indexing is incomplete.

### 4.2.2    Policies for Substance Indexing

The applications of many of the policies for indexing of substances are best illustrated through examples. A number of categories are overviewed in Table 4.1.

Built into SciFinder are algorithms that automatically allow for interpretation of many of these policies and that automatically handle issues of resonance, tautomerism, and substances which may be represented in different ways, such as open and ring forms of carbohydrates and pentavalent phosphorous halides. However, it pays for the scientist to consider why particular answers have been retrieved and if in doubt about what is occurring to seek explanations. In the sciences, the keen *observer* makes significant discoveries and this is also true with information retrieval.

---

*SciFinder Comment*

Understanding answers from substance searches requires understanding of two key aspects: issues with indexing substances (i.e. policies to represent structures in electronic databases) and SciFinder search algorithms (i.e. how SciFinder interprets the query).

In the rare cases where 'expected' substances are not retrieved, it is advisable to look for related substances, to check their registrations, and then to try alternative searches.

---

**Table 4.1**    *Overview of principal indexing issues for substances in REGISTRY*

| Issue | Summary of general indexing | Example |
|---|---|---|
| σ-bonds (where both bond electrons are provided by one of the atoms in the bond) | Represented as a double bond between the atoms | Appendix A4.1.5 |
| π-bonds | Represented as a single bond between all the participating atoms | Appendix A4.3.3 |
| Stereoisomers | Represented by stereochemical descriptors in the name field and in the structure | Appendix A4.1.4 |
| Isotopes | Hydrogen isotopes represented in the formula field; all isotopes represented in the name and structure fields | Appendix A4.1.3 |
| Resonance | Special bond description ('normalized bond') | Appendix 5 and Section 5.2.3 |
| Tautomerism | Individual valence bond structures of the tautomers are entered | Appendix 5 and Section 5.2.4 |
| Alloys | When precise ratios of elements are known, the ratios are listed in the name and composition fields | Appendix A4.2.2 |
| Salts | Generally indexed as two component substances with the acid and the base as separate components | Appendix A4.2.1 |
| Mixtures, hydrates, host–guest complexes | Indexed as substances containing separate components | Appendix A4.2 |
| Metal complexes | A variety of registrations depending on the nature of the complex | Appendix A4.3 |
| Polymers | A variety of registrations but mainly as starting materials (separate components for the monomers) or as products (where the polymer has a precise structure repeating unit) | Appendix A4.4 and Section 6.10 |
| Peptides/proteins | Peptides have sequence data; those with < 50 residues also have structure data | Appendix A4.4.4 and Section 6.9 |
| Nucleic acids | Nucleic acids have sequence data; those with < 5 residues also have structure data | Appendix A4.4.5 and Section 6.9 |
| Incompletely defined substances | A variety of registrations depending on the nature of the substance | Appendix A4.5.1 |
| Minerals | Compositions, where known, are given in the name and composition fields; otherwise indexed by name only | Appendix A4.5.2 |
| Natural oils, fats, etc. | Described by common or trade name, or by source in name and definition fields; in general they should be searched in CAPLUS by CA Index Name rather than by CAS Registry Number | Appendix A4.5.3 |

## 4.3   Searching for Substances: The Alternatives

Currently there are four main ways to search for substances in SciFinder. Sometimes the easiest way for those not familiar with the complexities of chemical substances is simply to search under **Explore: Research Topic** (Section 4.7).

The other three ways all start with **Explore Substances** and use either structure, name, or molecular formula based terms (Figure 4.1). Once the substances are found, the user clicks **Get References** and obtains answers in CAPLUS and MEDLINE. *The search terms used for the substances in this crossover from REGISTRY are CAS Registry Numbers* (linked to CAS Roles if requested), but as noted previously this is exactly what is required in nearly all cases.

*Searching by name* (Section 4.5) usually requires an exact match of the entry with a complete name of a substance in the database. It is the method of choice if a common or trivial name is known, and substances such as cholesterol, penicillin G, morphine, sodium acetate, and acetylene are easily found by this method. Indeed, most of the substances in biology, engineering, physics, material sciences, and pharmaceutical and

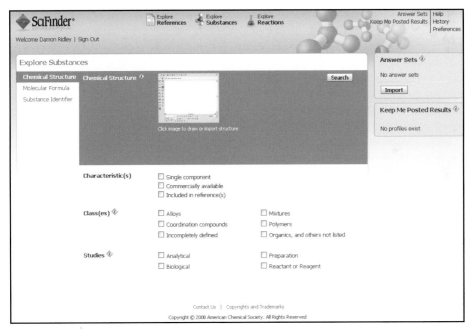

*Figure 4.1   Chemical Structure search screen. The Chemical Structure, Molecular Formula, and Substance Identifier options are on the left; additional options are below. Clicking the image activates the structure editor. SciFinder® screens are reproduced with permission of Chemical Abstracts Service (CAS), a division of the American Chemical Society*

medical sciences have relatively simple names, and scientists working in these fields will usually find the substances they commonly require through a name search.

However, it may be a challenge to find by name the specialty substances used in research chemical, biochemical, and pharmacological laboratories. Even simple substances like the chloroaldehyde (**1**) may be tricky to find by name-based terms,[1] and in any case the scientist would need to consider how the four possible stereoisomers of substance (**1**) would be named. Probably a structure search would be the easiest solution in this case!

(**1**)

*Searching by molecular formula* (Section 4.6) may require some knowledge of how formulas are entered, but even scientists with just a basic knowledge of substances may easily calculate formulas. However, many substances may share the same formula, and around 160 substances are found in a search on the molecular formula (C5 H7 Cl O) for the chloroaldehyde (**1**). These may take a while to look through, and probably the searcher would find it easiest and quickest to refine the substances with a structure search, so again knowledge of structure searching is valuable.

*Searching by structure* (Section 4.4) may require some knowledge of how substances are represented in the database, but once a few rules are understood then in the majority of cases structure searches will be the method of choice. Structure searches also open up possibilities to find substances related by structure (Chapter 5), but this important aspect is not possible through name or formula searches.

In the case of substance (**1**), it is a simple matter in SciFinder to draw and perform an **Exact search** on the structure. All stereoisomers and isotopically substituted substances are retrieved and the required answers may be selected.

## 4.4   Explore Substances: Chemical Structure

### 4.4.1   Overview

After the structure editor thumbnail is clicked (see Figure 4.1), the structure editor screen (Figure 4.2) appears and the user can draw a structure in a manner similar to that used in most computer chemical structure drawing programmes.

Next the user clicks an option below **Get substances that match your query**. No matter which option is chosen, the user is offered additional options. Once the user clicks **Search**, answers are obtained. But what types of answers are produced for each type of search? If **Substructure search** is chosen, the screen (Figure 4.3) is shown and choices are made. The structure search is then performed and answers (i.e. substances from REGISTRY) are obtained.

---

[1] If 4-chloro-2-pentenal is searched as a name, then SciFinder finds over 30 'possibilities', of which one is substance (**1**) (CAS Registry Number 107951-30-4) where the stereochemistry is undefined.

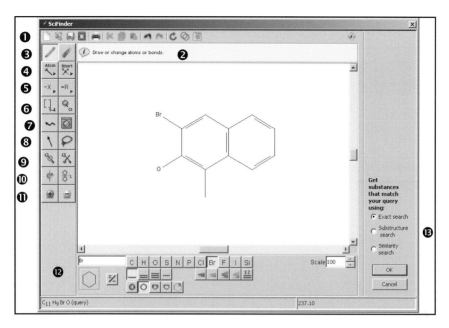

**Figure 4.2** *SciFinder's structure editor. A description of the numbered functions is given in Table 4.2. SciFinder® screens are reproduced with permission of Chemical Abstracts Service (CAS), a division of the American Chemical Society*

Many of the issues that may be encountered, and the differences between an **Exact** and **Substructure search**, are best illustrated through an example. Thus, suppose information on the primary amine represented in part by structure (**2**) is required. While there are a number of issues to consider before the search is commenced (e.g. stereochemistry?, exact substance?, salts?, substructures?), nevertheless suppose that the structure (**2**) is drawn and an **Exact search** is chosen. When this is done one answer with stereochemistry undefined (CAS Registry Number 878661-56-4) is obtained. But what has been searched and what should the user do next?

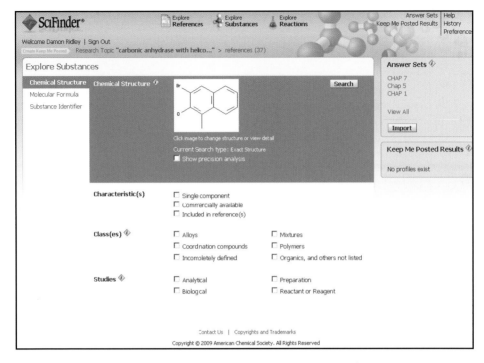

**Figure 4.3**   *Screen obtained after* **Substructure search** *is chosen from Figure 4.2. Query structure now appears in the structure editor thumbnail. SciFinder® screens are reproduced with permission of Chemical Abstracts Service (CAS), a division of the American Chemical Society*

The user may then think it is valuable to find substances related by structure, whereupon **Substructure search** is chosen. Over 1500 substances are found including substances with structures represented by structures (**3**) to (**7**). Why were some of these obtained? What is the next step?

It is apparent there are a number of issues; those relating to **Exact search** are discussed below. Further issues relating to **Substructure** and **Similarity searches** are discussed in Chapter 5.

### 4.4.2   Drawing Structures

When first using SciFinder, the user is advised to scroll through the pull-down menus at the top and the structure drawing tools on the left of the screen (Figure 4.2). Most of the options are self-explanatory and a summary is shown in Table 4.2.

Even from a glance at the functions in Table 4.2 it is apparent that considerable variation may be built into structure queries. The challenge is to think carefully at the outset what type of query needs to be built. In general it is better to build simpler structures and then to examine the types of answers retrieved, particularly since SciFinder has many additional analysis and refinement functions that may be used to narrow answers from structure searches at a later stage (Chapter 5).

**Table 4.2** *Brief description of functions displayed in Figure 4.2*

| Locator in Figure 4.2 | Function | Brief description |
|---|---|---|
| ❶ | Command options including New, Open . . . Erase | Implement normal file processing commands. |
| ❷ | Check overlaps. | Alerts users to overlapping atoms or bonds in the structure. |
| ❸ | Pencil tool (drawing). | Draws atoms/bonds. |
| | Eraser tool. | Erases atoms/bonds. |
| ❹ | Atom | Opens the Periodic Table of elements from which atoms may be chosen. |
| | Shortcut | Opens a number of defined groups of atoms, e.g. $CH_3$, $CH_2$, COOH, $NO_2$, etc., which may be chosen and quickly entered into structures. |
| ❺ | -X | Opens predefined variable groups, e.g. X (halogens), Ak (alkyl chains), etc., which may be chosen and quickly entered into structures. |
| | -R | Allows the user to define variable groups, e.g. a node may be any of C, N, P, which may be chosen and quickly entered into structures. |
| ❻ | Repeat | Allows a node or group, e.g. –OCH2CH2–, to be repeated any number of times between 1 and 20. |
| | Variable position of substitution | Allows a node or group to be substituted at variable positions in a ring. |
| ❼ | Chain tool | Allow chains to be drawn. |
| | Template tool | Allows rings from templates to be drawn. |
| ❽ | Selection tool | Allows node or bond to be selected and moved. |
| | Lasso tool | Allows parts of structures to be selected and moved. |
| ❾ | Atom lock tool | Stops substitution of non-H groups on atoms. |
| | Ring/chain lock tool | Stops (further) rings from being present in answers. |
| ❿ | Rotate tool | Allows selected groups to be rotated or flipped. |
| | Flip tool | |
| ⓫ | Charge tool | Allows either positive or negative charges to be present. |
| ⓬ | Common atoms | Allows common atoms/bonds to be drawn. Note the |
| | Common bonds | current atom and the current bond appear to the left |
| | Stereo bonds | of the list. |
| ⓭ | **Get Substances** | Allows various types of structure searches to be performed. |

*SciFinder Note*

Structures may be imported into SciFinder from other structure drawing programmes, but the structure drawing editor in SciFinder contains all the functions needed. In any case, the challenging aspects are related to deciding what structures to build and then working through structure search and post-processing applications – rather than the mechanics of structure drawing.

SciFinder also offers a quick way to import structures from REGISTRY records. Thus, after a structure in a substance record is clicked, a dialog box appears with two options: Explore by Chemical Structure and Explore Reactions. If the first is chosen, SciFinder then pastes the structure into the structure drawing editor; further modifications if needed can then be made in the usual way.

### 4.4.3  Explore Substances:  Exact search

There are three structure search options: **Exact search, Substructure search**, and **Similarity search**. The choice made depends on the intention of the search; some of the issues have been mentioned already in reference to structures (**2**) to (**7**). What is important is that the scientist understands the differences and applies the options that not only will best solve the immediate problem but also will lead to new discoveries.

In drawing the structure query it is not necessary to assign hydrogen atoms, since the exact search process automatically inserts hydrogen atoms at vacant positions. Further, while variable bonds (e.g. unspecified bonds) are allowed, variable atoms (e.g. generic groups Q, A, X, M, or R-groups) are not allowed in the query for **Exact search**.

Answers include all stereoisomers and isotopic substances, and multicomponent substances where the exact structure is one of the components. Examination of entries in Appendix 4 will suggest to users when multicomponent substances are important in the search. For example, many biologically active compounds are either organic acids or bases that are insoluble in biological fluids, and one of the key considerations in medicinal chemistry is to develop biologically compatible derivatives. This may be achieved either through the formation of salts or through complexation (e.g. with substances such as cyclodextrins). Since salts and supramolecular complexes are indexed as multicomponent substances with the biologically active substance as one component, it is apparent that an **Exact search** will retrieve these additional answers, which are of interest particularly for scientists in biological and medicinal fields.

Exact searches nearly always proceed to completion and the user does not need to be concerned with any of the additional issues involved with substructure searches (Chapter 5). However, if the query contains a structure that is present as a component in many multicomponent substances (e.g. styrene (**8**) is a monomer in almost 80,000 polymers) then large answer sets may be retrieved. In these cases, before the search is conducted some of the boxes in the **Characteristic(s)** and **Class(es)**, shown in the bottom of Figure 4.3, may be chosen. For example, if the **Characteristic: Single component** is chosen, **Exact search** on the styrene structure gives around 125 substances, some of which are shown in Figure 4.4. Note that these substances have been sorted by 'Number of References'. This allows the most common substances (in this case polystyrene and styrene) to be identified readily.

(**8**)

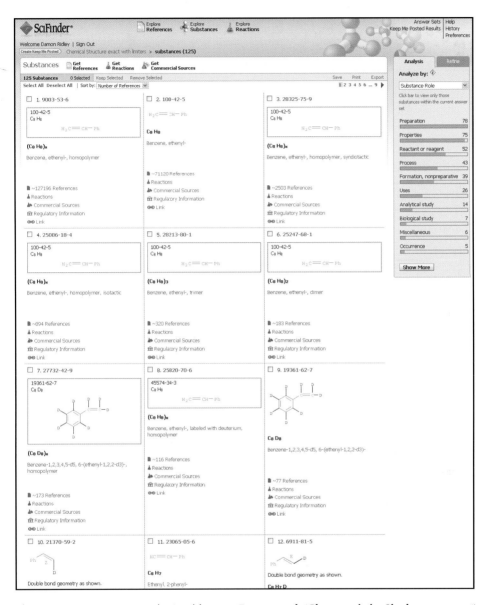

**Figure 4.4**    *Some answers obtained from an **Exact search** (**Characteristic: Single component**) on styrene (**8**). Isotopic substances, stereoisomers, the dimer, and homopolymer are obtained. It is a simple matter to narrow these answers further in SciFinder. SciFinder® screens are reproduced with permission of Chemical Abstracts Service (CAS), a division of the American Chemical Society*

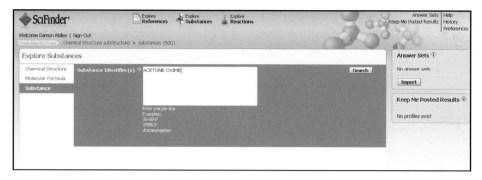

**Figure 4.5    Substance Identifier(s)** *screen. Names of substances or CAS Registry Numbers may be entered. SciFinder® screens are reproduced with permission of Chemical Abstracts Service (CAS), a division of the American Chemical Society*

## 4.5    Explore Substances: Substance Identifier

When **Explore Substance**s: **Substance Identifier** is chosen, the screen (Figure 4.5) appears, and names or CAS Registry Numbers may be entered. The search algorithm first looks for an exact match between the name entered and a complete name in the database. If an exact match is achieved the answer is displayed. If an exact match is not achieved, the algorithm attempts to give some reasonable answers by looking at parts of names.

Unless the user is proficient in nomenclature, name searches generally succeed best when common or trade names are known. However, even here there may be issues to address. For example, there are around 20 records for 'calcium sulfate' and its hydrates, and a number of these substances may be of interest. It is important that the searcher tries a few alternatives and if the required substance is not obtained in the first search, some of the answers may suggest how closely related substances may be named.

Matching names in the substance database is a challenge, and users familiar with the complexities of searching by name fragments will appreciate the difficulty in writing an algorithm to search for all name possibilities. In summary, SciFinder performs very well when exact matches are achieved, but in other cases the user may need to explore alternatives. Often it is a matter of exploring a few name-based possibilities; if they fail or if answers do not retrieve all required derivatives (salts, hydrates, etc.), then alternative searches described below need to be explored.

## 4.6    Explore Substances: Molecular Formula

When **Explore Substances: Molecular Formula** is chosen, the screen in Figure 4.6 appears and the formula is entered. Even though there are a number of rules relating to the order in which the elements appear in actual records (e.g. see the many examples in Appendix 4), the order in which the elements is entered is not important. SciFinder automatically arranges the elements in the required order and thus any of the entries

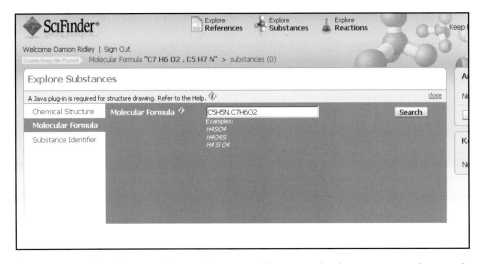

**Figure 4.6** *Molecular Formula search screen. The example shows an entry for a multi-component substance (molecular formula for pyridinium tosylate). SciFinder® screens are reproduced with permission of Chemical Abstracts Service (CAS), a division of the American Chemical Society*

C H2 Cl Br, H2 C Br Cl, or Br Cl H2 C will retrieve bromochloromethane, which in the database is represented by the molecular formula C H2 Br Cl.

In a similar way, to retrieve pyridinium benzoate (**9**), it does not matter whether the formula query entered is C5 H5 N. C7 H6 O2 or C7 H6 O2. C5 H5 N. Further, it does not matter in which order the components are entered in order to retrieve the polymer involving styrene, butadiene, and acrylonitrile (ABS; Appendix 4, Section A4.4.2). What does matter, however, is that the user understands that salts are indexed as components involving the acid and the base, and that the formulas for polymers are enclosed in parentheses followed by the suffix 'x'. For example, an entry to search by molecular formula for ABS may be (C8H8. C3H3N. C4H6)x.

(**9**)

One of the simplest ways to understand how molecular formulas are presented in the database is to examine records. Appendix 4 gives a number of examples and Table 4.3 gives a summary of the most commonly encountered issues.

An issue encountered is that many different substances have the same formula, so many substances may be retrieved in a single search. For example, a search on formula C10 H18 O2 gives over 6000 answers, but they may be refined either manually or by options described in Section 5.3.

***Table 4.3***   *Examples of representations of molecular formulas*

| Substance class | Molecular formula | Examples |
|---|---|---|
| Single component substances not containing carbon | Elements are entered in alphabetical order. | Sulfuric acid: H2 O4 S |
| Single component substances containing carbon | Carbon first, then hydrogen (if present) followed by elements in alphabetical order. | Cholesterol: C27 H46 O |
| Multicomponent substances (includes mixtures, hydrates, salts, inclusion compounds) | The individual components are represented using the rules above and components are placed in order, with those containing greatest number of carbons first. Further rankings (if needed) are based on numbers of hydrogens, followed by alphabetical listings of elements. | Calcium phosphate: Ca. 2/3 H3 O4 P (Also Ca. x H3 O4 P) Pyridinium tosylate: C7 H8 O3 S. C5 H5 N |
| Polymers | Component formulas are put in parentheses and 'x' suffix added. | Styrene/butadiene copolymer: (C8 H8. C4 H6)x |
| Coordination compounds | Individual compounds or ions follow rules for single component substances. | Cisplatin: Cl2 H6 N2 Pt |
| Alloys | Atoms in alphabetical order. | C . Cr . Fe . Mn . Ni . P . S |
| Isotopic substances | Only isotopes of hydrogen (D, T) have special listing in formula field. Isotopes of other atoms list symbol for atom only. | Deuterochloroform: C Cl3 D $^{13}$C acetic acid: C2 H4 O2 |

### 4.6.1   Examples of Applications of Searches by Molecular Formula

Searches on formulas may provide answer sets not possible by other methods. The following examples illustrate the power of formula searches.

First, if all possible trichlorobiphenyls (**10**) are required, an easy way to proceed is to search for all substances with formula C12 H7 Cl3. In this case almost 60 substances are retrieved; **Refine** (substances) with the biphenyl structure gives just over 40 trichlorobiphenyls, with different ring and isotopic substitutions.

(**10**)

(**11**)

(**12**)

Second, isotopes of hydrogen may be searched with deuterium (D) and tritium (T) in the formula (but isotopes of the other elements may be searched in the formula field only through the element symbol). The required substance then needs to be selected from the list of substances retrieved. Accordingly, a quick way to find the deuterated compound shown in Appendix 4, Section A4.1.3 is through a molecular formula search for C10 H13 D N2.

Finally, formula searches may be the preferred starting option for retrieval of reactive intermediates like carbocations, carbanions, and radicals. For example, one way to find the cyclopentylmethyl radical (**11**) is first to search for the formula C6 H11 and then to refine substances with the structure query drawn as shown in structure (**12**).

## 4.7   Explore References: Research Topic

As mentioned previously (Section 3.3.5), if one of the concepts detected in **Explore References: Research Topic** matches the name of a substance in REGISTRY, then SciFinder automatically searches the CAS Registry Number for the substance. **Explore References: Research Topic** also offers an option to look for substances, and even to find CAS Registry Numbers.

This process works better for bibliographic records in CAPLUS after 1984 when common names used by authors appear in many records. For example, the search for 'acetone oxime' through **Substance Identifier** (Figure 4.5) does not retrieve the substance (i.e. *this* name is not present in the REGISTRY database). However, **Explore References: Research Topic** 'acetone oxime' retrieves over 1000 bibliographic records 'as entered' and after looking through a couple of the records the CAS Registry Number (127-06-0) readily becomes apparent.

There are many other applications, for example:

- To find protonated methane (CH5+) it is a simple matter to enter **Explore References: Research Topic** 'CH5', which gives over 750 records 'as entered' and from these the CAS Registry Number (15135-49-6) is easily found (it is registered as a 'salt' with molecular formula CH4. H);
- Graft copolymers are often indexed as derivatives of the copolymer, so it is an easy matter to **Explore References: Research Topic** 'styrene butadiene with graft copolymer'. This gives over 200 records 'as entered' and from these records the indexing of various graft copolymers readily becomes apparent;
- One option to find preparations of isoxazoles through combinatorial methods is **Explore References: Research Topic** 'combinatorial with isoxazole or isoxazoles'. Approximately 60 references are found and once the indexing is understood the search can be broadened in many ways.

The general issue as to whether information on substances should be found directly through **Explore References: Research Topic** or through **Explore Substances** is discussed in Section 6.7, but for the present it is helpful to remember that SciFinder offers several different options. If the first option tried does not work, then it is a simple matter to try other options.

## 4.8   Summary of Key Points

- While most substances are registered in a way that scientists readily recognize, there are many cases where special registrations are needed for substances databases;
- One class of special registration is multicomponent substances, and this registration applies in many cases such as salts, hydrates, and copolymers;
- CAS Registry Numbers are the systematic index terms for substances in CAS databases;
- CAS Registry Numbers may be found in several ways in SciFinder including searches by name, formula, and structure;
- There are three ways to search by structure: Exact, Substructure, and Similarity;
- Exact structure search queries do not allow any variations in atoms (although bonds may be unspecified);
- Answers from Exact structure searches do not allow any additional atoms (other than hydrogen or deuterium or tritium);
- Answers from Exact structure searches include isotopic substances, stereoisomers, and multicomponent substances where the structure query exactly matches one of the components;
- Searches based on names of substances usually require exact matches with actual names in the database;
- SciFinder algorithms allow for variations in the order in which atoms are entered in searches based on molecular formulas, although it is preferable to enter formulas using REGISTRY conventions;
   - General policies are given in Table 4.3 and some examples are given in Appendix 4.
- Names of substances can be searched through **Explore References: Research Topic** and CAS Registry Numbers may be found by looking through actual answers.

# 5

# Substructure and Similarity Searching

## 5.1 Introduction

In structure searches, valence bond connections in queries are matched with valence bond connections in answers. In the **Exact search** process (Section 4.4.3), 'what you see is what you get'. There are few issues to address. However, it is more complicated with **Substructure** or **Similarity** searches where it helps to understand a number of issues.

The structure query is built first. Unlike for **Exact search**, the **Substructure/Similarity** query may have many variations relating to the types of atoms and bonds, and in some cases how they are connected (e.g. variable points of attachment). While considerable variation is allowed in the query, it usually pays to keep the structure relatively simple initially and to make modifications based on the substances obtained.

Next the query is searched in one of two ways:

- In a **Substructure search**, SciFinder retrieves all substances obtained through an **Exact search** plus substances with additional substituents where allowed on the structure query. The connectivities in the query are always preserved in the answers;

  – Substructure searches are used when answers with a definite part structure are required.

- In a **Similarity search**, the connectivities in the query are preserved in part, but not overall; i.e. answers from similarity searches will have recognizable parts of the query structure but the parts may be put together in ways that are driven by a number of computer algorithms. The algorithms operate at various levels of similarity, and answers are grouped from most similar through to least similar. The user looks through the different groups in turn and chooses answers of interest;

  – Similarity searches are used when greater variation in structures is of interest.

*Information Retrieval: SciFinder®, Second Edition* Damon D. Ridley
© 2009 John Wiley & Sons, Ltd

## 5.2    Searching Substances: Substructure

Many additional issues arise in the substructure search process.

### 5.2.1    The Screening Issue

SciFinder performs structure searches in steps. First, substances are screened; second, candidate answers are checked through atom-to-atom matches with the query; third, acceptable answers are sorted; and finally the answers are displayed.

The screening process mainly identifies small structure fragments, some atom connectivities, and various bond and ring types. SciFinder automatically generates screens for the query and then matches these with screens for substances in the database. If the number and type of screens match, then SciFinder pulls out candidate answers.

Candidate answers are then compared, atom-to-atom, to the query structure. There are two main reasons for this process. First, most screens have more than one fragment definition and the fragment in the query, which caused selection of that screen number, might have been different from the related fragment in the candidate answer. For example, the query might have had the atom sequence O–C–C–C–C–O whereas the potential answer might have had the atom sequence O–C–C–C–C–S (both of these fragments are defined by the same screen number).

Second, the different fragments that caused selection of the different screens might have been connected differently in the query and in the candidate answer. This is illustrated by part-structures (**1**) and (**2**), both of which contain two fragments defined by the screen number, but their connections are very different.

(1)                                                        (2)

### 5.2.2    Structure Is Too General

If the query is very general then too many *potential* answers pass the screening process and the query needs to be modified. When this occurs, SciFinder advises that the search could not be performed and gives a warning message: 'Structure is too general. Select limiter(s) below, draw additional atoms and bonds, or lock out rings and chains (Figure 5.1).

Each of these suggested options adds more screens to the search query, but adding more screens may not necessarily mean the search will complete. It all depends on the frequency with which the screens occur in substances in the database. Thus, for example, clicking the **Single component** limiter alone may make little difference because over 80% of the structure-searchable substances in the database have single components. There is of course no way of knowing in advance which screens have a large numbers of substances (although experienced chemists may make educated guesses); it really is a matter of trying a few options and observing the outcomes. The important thing is that restrictions that may exclude desired answers are not applied.

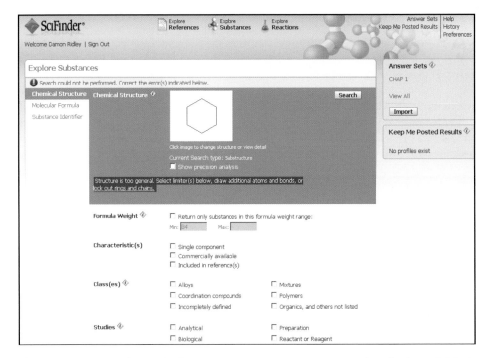

**Figure 5.1** *Screen obtained when a structure query is too general and where SciFinder is unable to complete the search. SciFinder® screens are reproduced with permission of Chemical Abstracts Service (CAS), a division of the American Chemical Society*

One of the most effective ways of applying screens that help the search to complete is to use the **Ring/Chain Lock Tool** because this not only helps to specify bonds in queries more precisely but also applies quite specific screens that are based on types of rings. For example, a search for substances with the substructure (**3**) will not complete (even though there are very few substances in the database with this substructure) unless the rings are locked; this process inserts very specific three- and four-member ring screens.

(**3**)

---

*SciFinder Tip*

The warning message 'Structure is too general' relates to the screen and search stages, and not to the answer stage. The problem is mainly due to the fact that the query has too few atoms or atoms in structures that are very common.

The addition of hydrogen atoms in the query may impact on the number of answers, but will have little impact on the screening process (since, with few exceptions, screens do not specify hydrogen attachments). Therefore other options discussed in this chapter need to be tried.

### 5.2.3   The Resonance Issue

Another issue relates to what constitutes an 'answer' and one consideration is the application of the concept of resonance in valence bond theory (Appendix 5). For example, if a **Substructure search** is performed on structure (**4**) then answers may include substituted furans (**9**), but the user may have wanted answers only where the ring in structure (**4**) is saturated (i.e. a tetrahydrofuran derivative).

   SciFinder usually interprets queries in the more general way, so if a substructure search on structure (**4**) is requested and if the ring is not locked, then SciFinder needs to allow for structures (**5**), (**6**), and (**7**). Consistent with the policy to provide more comprehensive answer sets and to allow the user to make evaluations in potentially marginal cases, SciFinder also allows for structure (**8**). However, since bonds in resonance structures are defined differently from single or double bonds (Appendix 5), SciFinder has to make a generic definition for the bond in the query (**4**) to allow for all possibilities. As a second six-membered ring may be fused on to (**8**) (a dibenzofuran structure), then in turn furans (**9**) need to be considered. If the ring is locked, then SciFinder does not have to allow for (**8**), and hence can define single bonds in the query and tetrahydrofuran structures result.

(4)          (5)          (6)

(7)          (8)          (9)

---

*SciFinder Tip*

Alternating single and double bonds in even-member rings are defined as 'normalized', i.e. the bonds are defined in a way different from single and double bonds. If rings in queries are not locked, then SciFinder may need to assign normalized/single or normalized/double bonds to the query. The outcome is that unsaturated substances may be presented in answers.

   If answers include structures that do not meet requirements then one solution is to use **Show precision analysis** (Section 5.2.5).

---

### 5.2.4   The Tautomerism Issue

Complications also arise with tautomers (Appendix 5). For example, if a **Substructure search** is requested for structure (**10**) then a possible answer is structure (**11**), which is a tautomer of structure (**12**). It could be argued that structure (**12**) contains the substructure (**10**) but then how is structure (**12**) related to structure (**13**)? There clearly is a problem with answers of the type (**13**) to a query based on (**10**), and a first thought may be to exclude the complication created by the tautomers (**11**) and (**12**). Such an exclusion leads to the question of how to treat structure (**14**), which clearly has the substructure (**10**) and which is related to structure (**11**).

In a similar way, structure (**15**) contains the substructure (**10**), but the tautomer issue arises with structure (**16**), which leads to a query concerning structure (**17**). Here it may be argued on chemical grounds that structure (**15**) will mostly be in the enol form (**16**), but how about structure (**18**)?

In summary, the substructure search algorithm needs to consider substances based on structures (**13**) and (**17**) when interpreting a query (**10**), and tetrahydronaphthalene and naphthalene derivatives (i.e. substances of the types (**13**) and (**17**) respectively) are retrieved. While the user may eliminate unwanted answers manually, or by modification of the structure query (e.g. by adding hydrogen or other atoms at key positions), in fact SciFinder offers other options that solve the problem much more elegantly.

### 5.2.5 Show Precision Analysis

In addition to the algorithms that address issues of resonance and tautomerism, SciFinder addresses automatically other structure issues such as chain and ring forms of carbohydrates, donor bonds, bonds to metal atoms in complexes and salts, and structures of

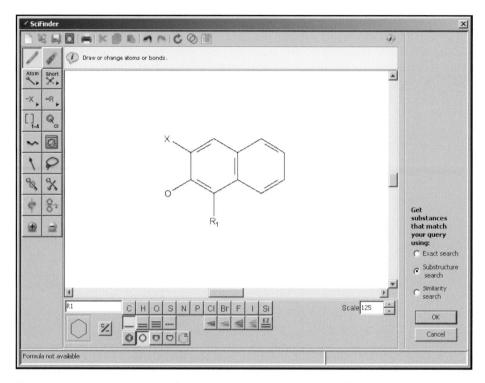

**Figure 5.2**   *Structure query. After the structure is drawn, one of three search options (on right) may be chosen. Precision Candidates screen for this query is shown in Figure 5.4. SciFinder*® *screens are reproduced with permission of Chemical Abstracts Service (CAS), a division of the American Chemical Society*

pentavalent phosphorus compounds. Other situations may arise where the user may not understand why certain answers have been retrieved; the reason probably is that SciFinder has applied some structure search algorithms automatically. The simplest way to check the interpretation of the query is through **Show precision analysis**.

If the structure shown in Figure 5.2 is built and if **Substructure search** is checked, the screen in Figure 5.3 appears.

After **Show precision analysis** is checked and then **Search** is clicked, the Precision Candidates screen is displayed (Figure 5.4). In this case SciFinder indicates that there are 758 substances that match the search query and another 277 substances that are (in some way) similar and that may be of interest.

If the box for Conventional Substructure is checked and then **Get Substances** is clicked, answers include substances (e.g. structures (**19**) and (**20**)) that contain the precise requirements of the query structure plus additional substituents at positions where substitution was allowed. Meanwhile structures (**21**) and (**22**) are included in the candidates 'Closely Associated Tautomers and Zwitterions' and 'Loosely Associated Tautomers and Zwitterions' respectively. In each case the part of the structure related to the query structure is highlighted.

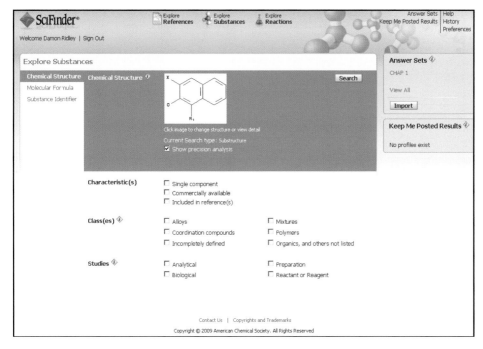

**Figure 5.3** *Structure search screen with **Show precision analysis** chosen. After **Search** is clicked, the Precision Candidates screen (Figure 5.4) appears. SciFinder® screens are reproduced with permission of Chemical Abstracts Service (CAS), a division of the American Chemical Society*

The user needs to understand how the sets are related to each other and to the query, and then to choose the set(s) that best meet the search requirements. Each case is different and users familiar with structural issues will soon understand how SciFinder has retrieved the different sets of answers. In the case of structures (**21**) and (**22**) the main issues relate to resonance/tautomerism, although it could be argued that (**21**) is a 'conventional substructure' match. It is precisely because the issues may be very subtle that the user should check the different sets of candidates.

**Figure 5.4**    *Precision Candidates screen for the query in Figure 5.2. It is advisable to check the different sets of candidates. SciFinder® screens are reproduced with permission of Chemical Abstracts Service (CAS), a division of the American Chemical Society*

---

*SciFinder Tip*

**Show precision analysis** is another tool that presents the user with options – before a decision needs to be made. It is usually best to check this tool routinely and then to work through the different sets. In many cases 'Conventional Substructure' will be chosen, but the other sets may alert the user to an unexpected issue, which may need to be explored.

---

### 5.2.6  Locking Tools

To understand further why some structures are retrieved it is helpful to know the *structure search defaults*. These are summarized in Table 5.1.

To use the locking tools, the appropriate tool (see ❾ in Figure 4.2) is clicked and then the cursor is clicked on the part of the structure to be modified. When an atom or bond is

**Table 5.1**    *Summary of **Substructure search** defaults*

| When the following is drawn … | The search default … | To override the default … |
|---|---|---|
| A ring | Is that the ring may be isolated or embedded in a larger ring system. | And retrieve only rings as drawn, use the **Ring/Chain Lock Tool**. |
| A chain | Is that the atoms drawn may be part of a chain or part of a ring. | And retrieve only atoms in chains as drawn, use the **Ring/Chain Lock Tool**. |
| An atom | In a **Substructure search** is that further substitution is allowed. | And block substitution, use the **Atom Lock Tool**. |
| A bond | Is that the bond (e.g. single, double, or triple) is searched as specified, although normalized bonds may be searched also. | View alternatives through **Show precision analysis** and choose required options. |

chosen, the **Ring/Chain Lock Tool** applies to all atoms in the ring system drawn or in the chain, and the atoms and bonds involved are highlighted in the structure displayed.

Presently, it is not possible to prevent formation of rings just at one position. Thus, clicking the tool on any atom in a ring isolates the entire ring, and clicking the tool on any atom in a chain isolates the entire chain. However, separate rings and different chains may be locked independently.

The **Atom Lock Tool** is used to block further substitution in a substructure search at the position locked. For example, if a five-carbon chain is drawn and the locking tool is clicked on the three middle carbons then structure (**23**) results. Further substitution at these carbons is thus prevented, and the search result is the same as if structure (**24**) or (**25**) had been drawn. Note that unless the ring/chain default option is overridden then cyclopentanes (**26**) will be retrieved in searches from any of these queries.

(**23**)          (**24**)          (**25**)          (**26**)

---

*SciFinder Tip*

Another use of the **Atom Lock Tool** is to control the oxidation states of atoms. For example, a **Substructure search** on query C–S–C retrieves sulfides, sulfoxides, and sulfones. The last two will not be retrieved if the sulfur is locked. If C–S(O)–C is drawn and the sulfur is locked, then a **Substructure search** retrieves sulfoxides only.

---

Particularly when used in initial queries, the impact of the **Ring/Chain Lock Tool** needs to be considered carefully. For example, consider possible answers (structures (**28**) to (**32**)) from a **Substructure search** on the query (**27**). If the ring was locked in the query then only structure (**29**) is retrieved. At least some of the other structures shown (particularly structure (**28**)) may have been of interest, but the user would not have been made aware of them.

---

*SciFinder Tip*

Research in science may be driven by the need to obtain a product or simply by curiosity. Either way, it is helpful if the scientist is always alert to new research possibilities, and through its ability to retrieve answers related to the initial query SciFinder may help considerably in research creativity.

Suppose that the motivation for doing a substructure search on structure (**27**) was that a new method for synthesis of these structures had been developed, or even needed to be developed. The more general search would have retrieved related substances, and it is possible that they may have become better research targets. It is best not to be too restrictive initially–the unexpected may be more interesting!

---

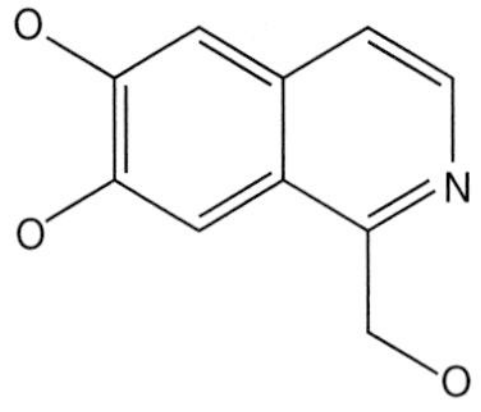

(27) 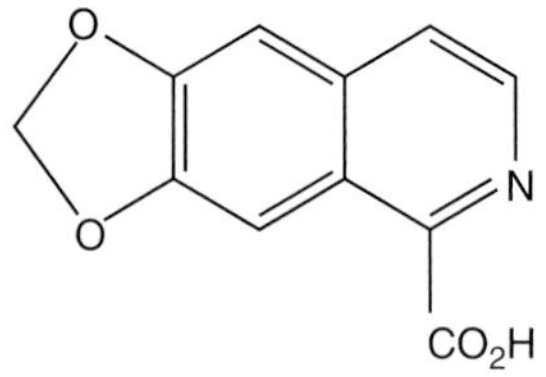    (28) 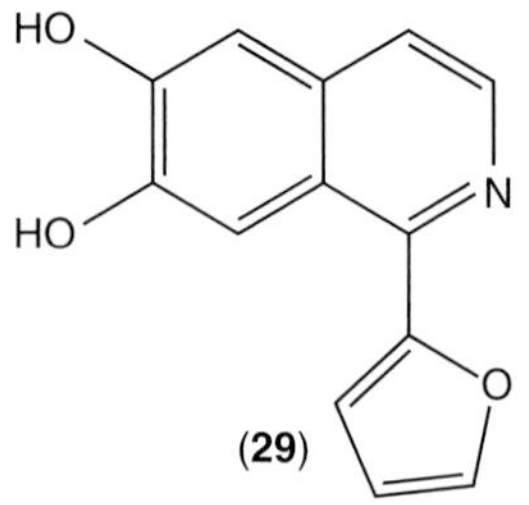    (29)

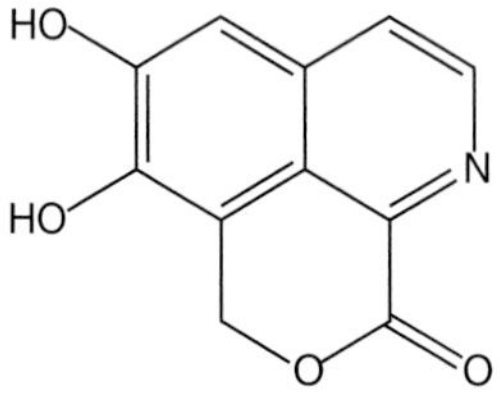

(30) 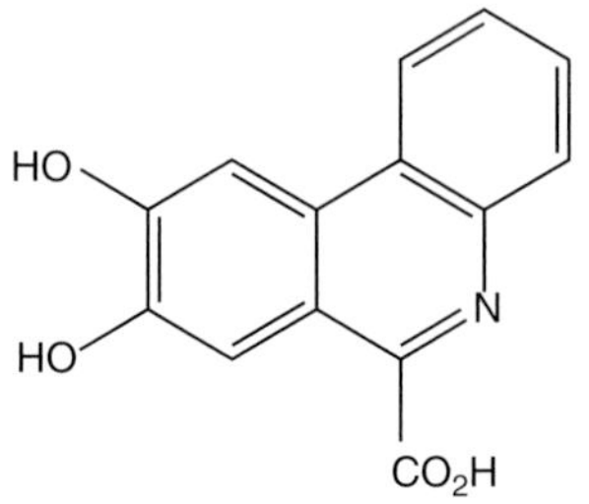    (31) 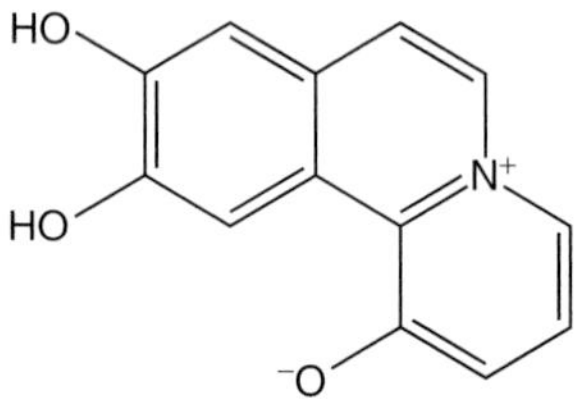    (32)

### 5.2.7    Additional Query Tools

The structure editor contains two additional tools (locator ⑥ in Figure 4.2) that apply to substructure search queries.

#### 5.2.7.1    *Repeating Group*

Atoms or groups of atoms may be repeated between 1 and 20 times with the **Repeating Group Tool**. Once clicked, the operation of this tool is explained in the yellow area located near the top of the structure editor. There are some restrictions on the use of this tool, e.g. at least one node must be included, stereo bonds are not included, and alkyl groups cannot be joined in the group (i.e. as Ak-Ak).

#### 5.2.7.2    *Variable Points of Attachment*

Atoms or groups of atoms may be attached at variable positions in a ring. First the atoms and the ring are drawn separately and once clicked the operation of the **Variable Points of Attachment Tool** is explained in the dialog box. Sometimes it is easier to check the positions of attachment if the **Lasso Tool** is used to highlight the connecting point, which then is dragged outside the ring (Figure 5.5).

The purpose of each of these tools is to allow specific variations in the structure. Note that it is usually necessary to make some restrictions on the atoms or groups of atoms. For example, if instead the repeating group shown in Figure 5.5 were simply O–C–C then a large number of answers would result, including esters O–C(O)–C. Restrictions may be quite specific, e.g. the repeating group in Figure 5.5 could be – O–CH(Me)–CH2.

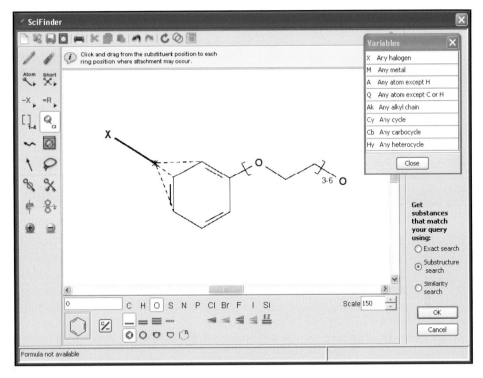

**Figure 5.5** *Screen showing structure with **Variable Point of Attachment** (VPA) and **Repeating group**. Instructions for use of the VPA tool and options for **Variables** are also shown. SciFinder® screens are reproduced with permission of Chemical Abstracts Service (CAS), a division of the American Chemical Society*

### 5.2.8   Additional Search Refinements

The search screen allows a number of additional requirements to be made (Figure 5.6). However, many of these modifications may be made once initial answers have been obtained (see **Analyze** and **Refine Substances** below).   Consistent with the general advice 'Let's see what we get first and then make decisions on how to proceed based on actual answers', it may be better not to check the **Characteristic(s)** or **Studies** boxes when *initial* searches are being performed.

However, this screen allows **Class(es)** of substances to be selected and it may be advisable to check appropriate entries. (This may also be done through **Refine: Chemical Structure**, but ultimately these types of modifications cannot be done in this way through any other option.) Examples of the **Class(es)** of compounds are given in Appendix 4, and in particular Sections A4.2.2 (Alloys), A4.2.3 (Mixtures), A4.3.1 (Coordination Compounds), A4.4 (Macromolecules), and A4.5.1 (Incompletely Defined Substances) should be noted.

| Characteristic(s) | ☐ Single component | |
| :--- | :--- | :--- |
| | ☐ Commercially available | |
| | ☐ Included in reference(s) | |
| **Class(es)** ◈ | ☐ Alloys | ☐ Mixtures |
| | ☐ Coordination compounds | ☐ Polymers |
| | ☐ Incompletely defined | ☐ Organics, and others not listed |
| **Studies** ◈ | ☐ Analytical | ☐ Preparation |
| | ☐ Biological | ☐ Reactant or Reagent |

**Figure 5.6**   *Additional requirements allowed in the initial structure search.* **Characteristic(s)** *and* **Class(es)** *options may be chosen now, if definitely known, but it may be advisable to choose options under* **Studies** *later (through* **Analyze: Substance Role** *since* **Analyze** *presents histograms that help to make decisions). SciFinder*® *screens are reproduced with permission of Chemical Abstracts Service (CAS), a division of the American Chemical Society*

## 5.3   Searching Structures: Working from the Initial Substance Answer Set

Once initial answers in REGISTRY are obtained there are many paths that may be followed, and one of the first may be to sort answers by **Number of References** (locator ⑥ in Figure 5.7). Sometimes it helps to look first through the substances with the most records since this may give a better idea of the important substances involved.

Other important links from this screen and brief descriptions are shown in Table 5.2. Note:

- Further information such as references, reactions, and commercial sources may be obtained for the entire answer set, for individual substances, or for selected substances;
- The information comes from the CAPLUS/MEDLINE, CASREACT, and CHEMCATS respectively;
- Additional refinements (e.g. Substance Role) are available because SciFinder algorithms use further indexing tools in the linked bibliographic databases.

### 5.3.1   Analysis of Substances

There are always two aspects to consider: how SciFinder implements the **Analyze** functions for substances and when to use them.

The key to the former is that CAS Registry Numbers are the systematic indexing terms for substances in CAPLUS, CASREACT, CHEMLIST, and CHEMCATS. Therefore, SciFinder simply checks which CAS Registry Numbers in the answer set appear in the other databases. In **Analyze: Substance Role**, SciFinder is further checking the CAS Roles associated with the substance(s). In the case of **Analyze: Elements**, SciFinder is looking through the Molecular Formula fields for the substances.

The main use of these functions is to provide the searcher with additional information before refinements need to be implemented. Remember that finding substances is only

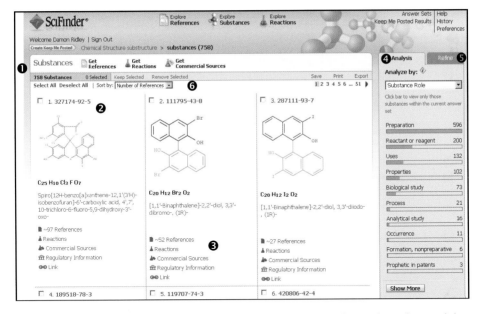

***Figure 5.7*** *Answer screen after 'Conventional Substructure' is chosen from the candidates in Figure 5.4 and after the initial list of answers is* **Sort***ed by* **Number of References***. The numbers in this figure are referred to in Table 5.2. SciFinder® screens are reproduced with permission of Chemical Abstracts Service (CAS), a division of the American Chemical Society*

the first step in the process and that the real goal usually is to find the original documents that report the information required.

The user needs to evaluate information as it is retrieved, and evaluations of answers may start by thinking: 'A search on the substructure Figure 5.2 leads to around 750 (conventional) substances (Figure 5.4). That's probably too many; however, it may be possible to narrow answers depending on the type of information required. Since my principal interest is the analysis of substances of this type, then **Analyze: Substance Role** indicates only 16 substances are involved (see Figure 5.7). That's a manageable number, so let's continue directly to narrow answers just to those which are indexed by Analytical study.'

Depending on requirements, other thought processes may be:

• 'I note there are only 27 substances that are commercially available, so a next step may be to focus on them, since I want to minimize my synthetic steps;'
• 'There are 596 substances that have preparation information, and that's too many. So, I probably should use some substance refinement tools to narrow substances first;'
• 'The initial query asked for any halogen and I get an idea (Figure 5.8) of the numbers involved from the histogram of elements. Perhaps I should focus first on the fluorine containing substances (98 substances);'
• 'I'm happy with the initial answer set and intend to narrow answers by **Explore References: Research Topic** in CAPLUS, so I'll proceed to **Get References**.'

**Table 5.2**   *Brief description of functions displayed in Figure 5.7*

| Locator in Figure 5.7 | Function | Brief description | For more details see Section |
|---|---|---|---|
| ❶ | **Get References** | Retrieves references for all substances. | 6.6 |
| | **Get Reactions** | Retrieves indexed reactions for all substances. | |
| | **Get Commercial Sources** | Retrieves only those substances that are commercially available. | |
| ❷ | CAS Registry Number | Links to the full record in REGISTRY. | 3.6 |
| ❸ | **References, Reactions, Commercial Sources, Regulatory Information** | Retrieves information for the substance only. | 6.6 |
| ❹ | **Analysis.** Options are:<br>• Commercial availability;<br>• Elements;<br>• Reaction availability;<br>• Substance role (the default) | Gives histogram of:<br>• substances that are (or are not) commercially available;<br>• elements present in substances;<br>• substances for which chemical reaction information in CASREACT is available;<br>• CAS Roles for the entire answer set. | 5.3.1 |
| ❺ | **Refine.** Options are:<br>• Chemical structure;<br>• Isotope-containing;<br>• Metal-containing;<br>• Commercial availability;<br>• Property availability;<br>• Reference availability;<br>• Experimental property;<br>• Atom attachment. | Retrieves substances of the type indicated.<br><br>Note that many of the options have choices, which are displayed once the option is clicked. | 5.3.2 |
| ❻ | **Sort.** Options are<br>• CAS Registry Numbers (the default);<br>• Number of references. | Sorts answers<br>• so that the latest entered substance appears first in the list;<br>• so that the substance with the most references appears first in the list. | 5.3 |

### 5.3.2   Refine Substances

Having considered the impact of **Analyze**, it may be decided to narrow substances through the seven options under **Refine**. Typical screens for the first six options are shown in Figure 5.9.

#### 5.3.2.1   Refine: Chemical Structure

To implement this function the structure image is clicked, the structure editor screen appears, and the query is modified. When finished, further refinements may be chosen

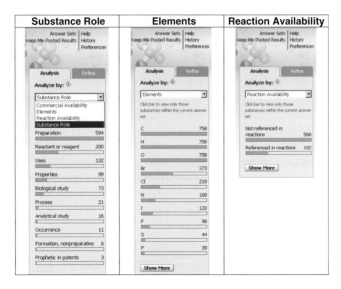

**Figure 5.8** *Analysis screens for the 758 answers obtained through **Substructure search** of query in Figure 5.2. Before any refinement of the substances is undertaken, it is advisable to review the types of information available on the substances. SciFinder® screens are reproduced with permission of Chemical Abstracts Service (CAS), a division of the American Chemical Society*

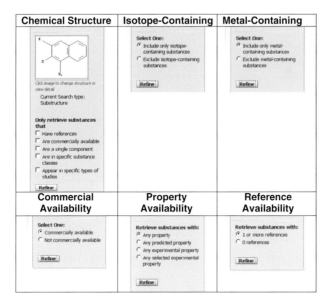

**Figure 5.9** *Refine screens for the 758 answers obtained through **Substructure search** of query in Figure 5.2. SciFinder® screens are reproduced with permission of Chemical Abstracts Service (CAS), a division of the American Chemical Society*

through options after **Only retrieve substances that** ... ; i.e. the full structure query and search tools are available.

However, the initial search should always be viewed as a first step only, and depending on the types of answers retrieved the user may immediately decide to go down an alternative path. For example, the substance with most references (Figure 5.7) may have stimulated new ideas and it may be of interest to start again with a substructure based on this structure. This is achieved most readily first by clicking on the CAS Registry Number 327174-92-5 (Figure 5.7), in which case the REGISTRY record (Figure 5.10) is obtained. Next, when the structure is clicked, two options are presented and **Explore by Chemical Structure** is chosen.

SciFinder then pastes the structure directly into the structure editor and subsequent changes to the structure may be made in the usual way (Figure 5.11). In this case, a **Substructure search** retrieves 15 substances for which references may be obtained.

### 5.3.2.2   Other Refine Options

The initial 758 substances may be refined to include or exclude isotopes, metal-containing substances, substances that are commercially available, and substances for which no references are available (Appendix 4, Section A4.5.3). SciFinder has screens for isotopic

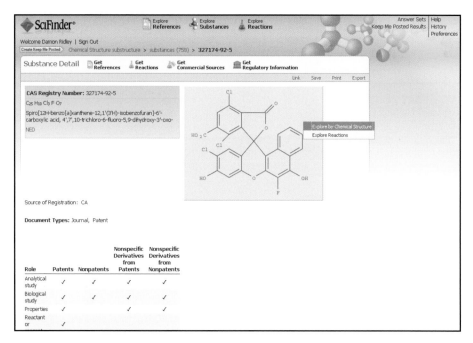

**Figure 5.10**   *When a substance in REGISTRY is clicked, two options appear.* **Explore by Chemical Structure** *inserts the structure into the structure editor and structure refinements may be made.* **Explore Reactions** *inserts the structure into the reaction editor, where structure and reaction refinements may be made. SciFinder*® *screens are reproduced with permission of Chemical Abstracts Service (CAS), a division of the American Chemical Society*

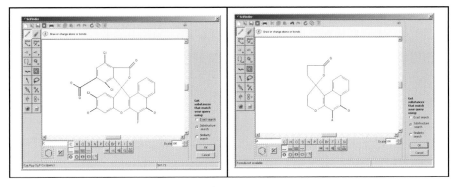

**Figure 5.11**   *Structure editor screens obtained before (left) and after (right) the structure is modified. SciFinder® screens are reproduced with permission of Chemical Abstracts Service (CAS), a division of the American Chemical Society*

**Figure 5.12**   *Options to **Refine Substances** to those with specific properties. SciFinder® screens are reproduced with permission of Chemical Abstracts Service (CAS), a division of the American Chemical Society*

and metal-containing substances and applies these in the first two cases. At present substances may be refined by broad property requirements (predicted and/or experimental properties) and by a number of specific properties (Figure 5.12), which become available after the box **Any selected experimental property** is clicked.

The remaining refinement option is **Atom Attachment**, and when this option is checked a new screen (Figure 5.13) appears. At first only the query structure is shown, but after an atom in the structure is clicked SciFinder displays a list of the atoms attached at this position. Options may be checked and so quite specific refinements may be made to the query.

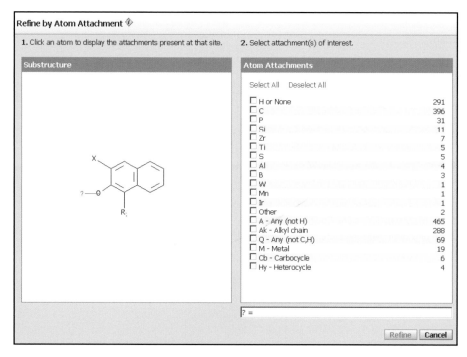

**Figure 5.13   Refine: Atom Attachment** *screen, which shows substituents in answers at positions left open in the query structure. In this example, the oxygen atom is clicked and atom attachments at this oxygen are shown in the right-hand column. SciFinder*® *screens are reproduced with permission of Chemical Abstracts Service (CAS), a division of the American Chemical Society*

Perhaps there is no better example of the power of SciFinder! Think how restrictive a search with a query with either a hydrogen or a carbon on the oxygen may have been. Further, what opportunities may have been missed? Through leaving the oxygen open and then later reviewing atom attachments, a number of options become apparent. Users should not overspecify initial queries!

### 5.3.3   Narrowing and Broadening Answers

**Analyze** and **Refine** offer many options to *narrow* answers, but even here the user needs to know what is being excluded. Some general issues include:

- *Get references.* Answers will be bibliographic records in CAPLUS and MEDLINE that have the CAS Registry Number as an index entry – other records may be of interest, e.g. very recently added records may not have full indexing at the time the search was undertaken;
- *Commercial availability.* Answers are limited to those listed in CHEMCATS. While this list is kept up-to-date as much as possible, some suppliers may not have provided their latest catalogues, and there may be other suppliers who have not sent their catalogues to CAS;

**Table 5.3**   *Options to broaden or narrow substructure searches*

| Broadening answers in substructure searches | Narrowing answers in substructure searches |
|---|---|
| Specify fewer atoms in structure. | Specify more atoms in the structure, i.e. build a more complex structure. |
| Leave as many positions open as possible. | Stop substitution at atoms as required. |
| Allow isolated/embedded rings (the default). | Isolate ring(s). |
| Allow chain atoms and bonds to have ring or chains values (the default). | Restrict atoms and bonds in chains to chain-only values. |
| Remove hydrogens. | Block substitution (e.g. by addition of hydrogens in vacant positions). |
| Use generic groups A, M, Q, X or allow atom variables (R-groups) rather than use specific aoms. | Define atoms more precisely (e.g. restrict generic groups). |
| Allow unspecified bonds. | Define bonds more precisely. |
| Use **Similarity search**. | Use advanced options: **Characteristic(s)** and **Class(es)** (Figure 4.1). |
| | Use options under **Analyze: Substances** as a guide to obtain more precise answers. |
| | Use **Refine: Substances** and thus retrieve a subset of the answers based on the part-structure drawn. |
| | Use **Refine: Substances** and limit answers by **Property Data**. |

(Note: a large *substance* answer set may be acceptable since information on the substances may appear in relatively few *references*. References may also be narrowed either through one of the CAS role filters or through **Analyze** or **Refine**.)

- *Reaction availability.* Coverage in CASREACT depends in part on the time period involved and in part on CASREACT indexing policies. For example, CAS has indexed all reactions from core organic synthesis journals over the last decade, and representative reactions before then. John Wiley's *Encyclopedia of Reagents for Organic Synthesis (EROS)*, *Organic Synthesis*, and *Organic Reactions* have been added to CASREACT recently, and have increased overall coverage, although again entry of reaction information in these sources is subject to indexing policies;
- *Metal-containing substances.* Silicon (Si) is not covered in this class.

While these issues may have impacts from time to time, the coverage of information and implementation of search tools in SciFinder remains unrivalled. On the other hand, options to *broaden* answers in substructure searches remain under the control of the user. Some of these are given in Table 5.3, together with a summary of options to narrow structures.

## 5.4   Similarity Search

Another way to broaden answers is to perform a **Similarity search**. The structure query is built in the same way as described in Section 4.4.2 although it is important to note that **Similarity search** does not support structures with variable groups (system- or user-defined), repeating groups, variable points of attachments, or multiple fragments, and stereo bonds in structures are ignored in searching.

Once the structure is built, SciFinder assigns a number of chemical descriptors (similar to the screens mentioned in Section 5.2.1) and then matches the descriptors with those for substances in the database. The number of query structure descriptors, file substance descriptors, and descriptors in common are then scored according to a formula that gives a rank for each substance in the database; SciFinder then displays a histogram of rankings.

Accordingly, after **Similarity search** is chosen for the query in Figure 5.14, the Similarity Candidates screen (Figure 5.15) is displayed. At this stage it is a simple matter to work through the different sets in turn and to examine them for relevancy.

In this case the exact substance (**33**) appears as the most similar candidate, while substances in the next set (90–94% similarity) are naphthols (**34**), (**35**), (**36**), and (**37**). The principle of **Similarity search** is thus demonstrated; the application depends very much on the query, and in practice it is necessary to look through a few of the ranked lists.

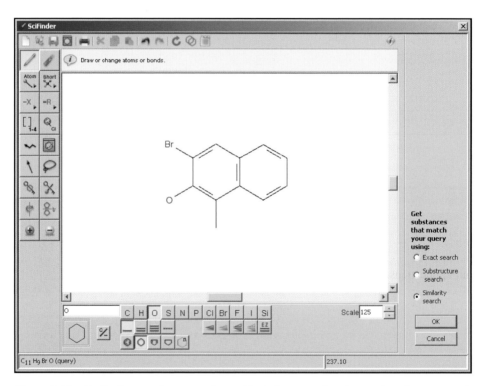

**Figure 5.14    Similarity search** *query. A specific halogen and carbon rather than a generic X node and R1 (Figure 5.2) are drawn since **Similarity search** queries do not allow variable atoms. SciFinder® screens are reproduced with permission of Chemical Abstracts Service (CAS), a division of the American Chemical Society*

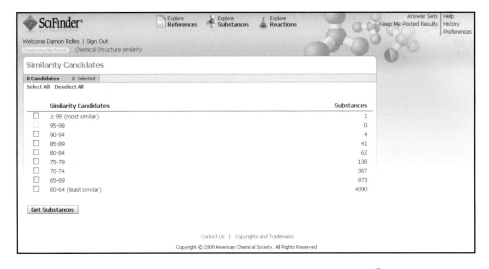

**Figure 5.15**   *Similarity Candidates for query in Figure 5.14. SciFinder® screens are reproduced with permission of Chemical Abstracts Service (CAS), a division of the American Chemical Society*

## 5.5   Further Examples of Show Precision Analysis

Section 5.2.5 gives an example of the application of **Show precision analysis** to the understanding of the resonance and tautomerism issues, but the function has other important applications.

### 5.5.1   Coordination Compounds and Salts

In the same way that organic chemists need to deal with tautomerism and resonance, coordination chemists need to deal with what constitutes a salt and what constitutes a coordination compound; i.e. if the structure query (**38**) is drawn, then would substances (**39**), (**40**), and (**41**) be reasonable answers from a substructure search? The point is that iron (III) acetate (**40**) is indexed as a salt whereas the other compounds are treated as coordination compounds (Appendix 4, Section A4.3), yet in one sense all contain the part structure (**38**). So when a **Substructure search** on query (**38**) is performed, SciFinder needs to allow for all possibilities and usually does so by effectively ignoring the bonds to the metal atoms (although this is not always the case as in some instances, e.g. metal-containing porphyrins, the algorithm does not ignore the bond to the metal). However, in being comprehensive, issues of precision again arise.

(38)          (39)          (40)          (41)

For a structure search query (**38**), the SciFinder solution is to place coordination compounds (e.g. substances (**39**) and (**41**)) in the precision candidates 'Conventional Substructure' and to place salts (e.g. iron (III) acetate (**40**)) in candidates 'Closely Associated Tautomers and Zwitterions'.

### 5.5.2    Cyclic Hemiacetals and Hydroxycarbonyls; Pentavalent Phosphorus

Another example is the case of hydroxycarbonyl compounds that may be in equilibrium with the cyclic forms (hemiacetals). Thus when an **Exact search** is conducted on query (**42**), currently six substances (the exact substance, isotopic substances, and an ionic substance) are retrieved under 'Conventional Exact' and another 34 under 'Closely Associated Tautomers or Zwitterions'. Two substances in this latter group are substances (**43**) and (**44**), which chemists recognize as the cyclic and enol forms of substance (**42**); most of the other substances in the latter group are multicomponent substances that contain substance (**43**) as one component.

(42)          (43)          (44)

(45)          (46)          (47)

Meanwhile, an **Exact search** on query (**45**) gives the substance as drawn, while **Show precision analysis** reveals two 'Closely Associated Tautomers or Zwitterions', namely substances (**46**) and (**47**).

These examples refer to **Show precision analysis** for an **Exact search**. The same applies to **Substructure search**, where the story may be a little more complicated because of additional substituents allowed on the query structure, although the principles remain.

*So When are Substances Indexed as keto or enol Forms, as Linear or Cyclic Hydroxycarbonyls, or as Salts or Complexes?*

The answer is that it depends mainly on what is presented in the original document. Thus if the authors represent dibromotriphenylphosphorane as substances (**45**), (**46**), or (**47**), then the document analyst reflects this in the indexing. If authors and referees agree on a representation, particularly in the context of the document, then it is not for the document analyst to enter an editorial interpretation!

## 5.6　Additional Structure Query Options

Structure queries may contain structure fragments, which opens up several possibilities.

### 5.6.1　Exact Search

If name or formula searches for multicomponent substances do not retrieve required answers, then structure searches involving fragments may be tried. Thus, if it is known that 'torpex' (Appendix 4, Section 4.2.3) is a mixture that contains aluminium, TNT, and trinitrohexahydrotriazine, it is a simple matter to draw the structures of these substances as shown in Figure 5.16. An **Exact search** then retrieves the substance.

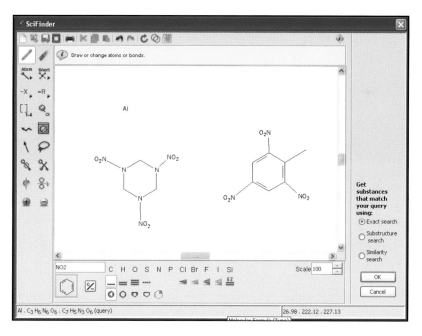

**Figure 5.16**　*Exact search for multicomponent substances. The query shows the structures of the individual components for Torpex. SciFinder*® *screens are reproduced with permission of Chemical Abstracts Service (CAS), a division of the American Chemical Society*

### 5.6.2  Substructure Search

If **Substructure search** is chosen for the query in Figure 5.16 then substances that contain at least these three components (and various substituted derivatives) are retrieved. Fragments are also used in substructure searches to retrieve substances with the part-structures, irrespective of how they are joined in a single component or in different components in multicomponent substances.

As an example of the former, if substances with two sulfolene groups are required then a **Substructure search** on the query (**48**) may be undertaken, and interesting substances of the types shown in structures (**49**), (**50**), and (**51**) are retrieved. It would be difficult to retrieve this variety of substances with any other query, and the result highlights the need to search more general queries initially. Users do not know what may turn up, although if too many irrelevant answers appear then it is an easy matter to refine them subsequently.

(**48**)

(**49**)                    (**50**)                    (**51**)

## 5.7  Getting References

While at times the substances alone are of interest, generally the *references* are required. In **Explore Substances** the answers retrieved are substances in REGISTRY, and when **Get References** is chosen the CAS Registry Numbers for the substances are used as search terms. The references obtained are from CAPLUS and MEDLINE. Then all the **Refine/Analyze/Categorize** options discussed for references in Section 3.5.5 and following sections or the **Get Substances/Reactions/Cited/Citing** options (Section 6.6) may be used in the normal way.

## 5.8  Combining Explore Substances and Explore References

The combination of searching for substances by name, formula, or structure in REGISTRY together with bibliographic and keyword information in CAPLUS and in MEDLINE offers unique and valuable opportunities for searching for information in the sciences through SciFinder. Additionally the **Analyze/Refine** functions for substances and the **Analyze/Refine/Categorize** functions for bibliographic records offer ways to narrow answers in systematic ways and to explore related science.

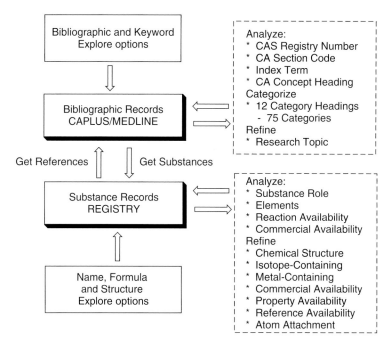

**Figure 5.17** *Summary of **Explore** and post-processing options in the bibliographic databases CAPLUS and MEDLINE and in the substance database REGISTRY. The **Get Substances** function is discussed in Section 6.6.1*

A summary of the functions is given in Figure 5.17. It remains for the scientist to take full advantage of these opportunities.

## 5.9   Summary of Key Points

- Sometimes the representation of chemical structures and chemical bonds is complex. It helps if searchers understand some of the structure conventions used *by CAS* in order to obtain good answer sets;
- Particular issues arise in many cases including:
  - aromatic compounds/resonance;
  - tautomers;
  - $\pi$-bonding;
  - donor bonding.
- SciFinder automatically interprets many of these issues, but at times more precise answer sets need to be obtained through **Show precision analysis**;
- In substructure searches SciFinder defaults to:
  - rings are isolated or fused;
  - chains are assigned chain or ring values.
- When the user chooses **Get References** from a structure answer set, SciFinder searches for the CAS Registry Numbers for the substances in CAPLUS and MEDLINE.

*Remember*

It is better to start out with a broader search and then refine answers rather than to start off too precisely. In this way, not only is the risk of missing key answers reduced, but also related information may be retrieved which in fact may prove more valuable than just that obtained through a very specific search. As a general rule, the user should start the search at a more general level; then if the initial answer set is too large the user may narrow the search with options under **Refine** or **Analyze**, or build more atoms into the structure, or just **Get References** for the large number of substances and **Refine/Analyze/Categorize** in the bibliographic databases.

# 6

# Additional Search and Display Options

## 6.1 Introduction

SciFinder offers many additional functions. Some of these are **Explore** options (**Author Name, Company Name, Document Identifier, Journal**, and **Patent**) which afford initial answer sets in the bibliographic databases, while others (**Get Substances, Get Reactions, Get Citing**, and **Get Cited**) are functions that are used after initial answers are obtained.

There also are a number of other issues to consider, including:

- When to use **Explore References: Research Topic** or **Explore Substances** to find substances;
- The complementary nature of the CAS and MEDLINE databases and when to take advantage of database specific features;
- How to search for other classes of substances, e.g. the substances in biology, and polymers.

These topics are discussed in this chapter.

## 6.2 Explore: Author Name

The exact entry of an author name in the original publication depends partly on how the *authors* wrote their names in the article and partly on the policies of the *editor* of the publication. Variations in representations occur particularly with entries for first names, when either full first names or just initials appear in the original document.

The situation is further influenced by *database producers*, who may apply additional policies. For example, whereas entries in CAPLUS are the same as in the original article,

*Information Retrieval: SciFinder®, Second Edition* Damon D. Ridley
© 2009 John Wiley & Sons, Ltd

up until recently entries in MEDLINE have been author last name and *initials only* (and since then entries have been the name as it appears in the original document).

Variations also may occur through the ways in which databases interpret phonetics (e.g. Müller is entered as Mueller), prefixes (e.g. van and de), and hyphenations. Names may be translated into American English (e.g. most Chinese language documents have author names in Chinese characters, but names that appear in the database are translated). Further, in some original documents an author's last and first names may be in different word order and names may be misspelt. Consequently, a single author may be represented in many ways, and the searcher usually needs to consider a number of options.

In SciFinder, authors are searched through **Explore: Author Name** (Figure 6.1). By default variations in spellings in last names but not in first names are searched. If initials are entered then candidates with possible full first names appear, while if full first names are entered candidates with the corresponding initials appear. Because of the possibility of many alternatives, the entry should be kept as short as possible (e.g. last name and first initial), although if the name and initial are quite common the list of candidates may be long and may take some time to work through.

The entry in Figure 6.1 produces 42 candidates (Figure 6.2), of which at least 15 are relevant, and indicates some alternative spellings for the last name, which in this case are few in number but still are relevant. It is difficult to construct an algorithm that retrieves all variants, but generally SciFinder helps considerably. If comprehensive results are required the user still may have to try a few alternatives to ensure that all records are identified.

Another issue relating to searching for authors is that different authors may have the same names, or at least the same last names and initials. The problem is often

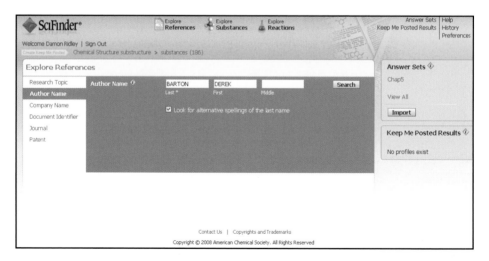

**Figure 6.1    *Explore: Author Name*** screen. In this case the full first name is entered (when the initial 'D' is entered over 100 options are presented). SciFinder® screens are reproduced with permission of Chemical Abstracts Service (CAS), a division of the American Chemical Society

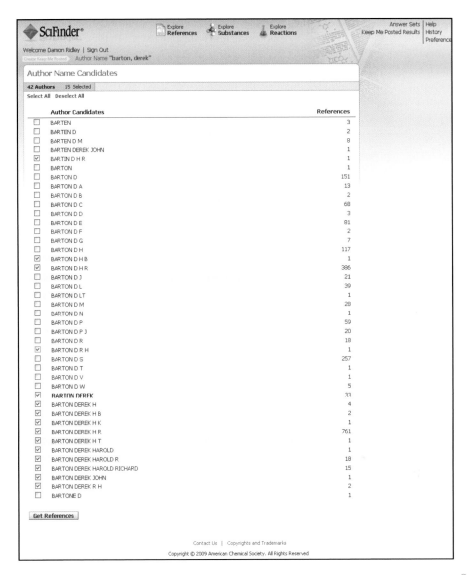

***Figure 6.2*** *List of candidates from **Explore: Author Name** shown in Figure 6.1. SciFinder®*
*screens are reproduced with permission of Chemical Abstracts Service (CAS), a division of*
*the American Chemical Society*

more complex for some names of Chinese/Korean/Taiwanese origin, where many authors
share the same last name and may have only one first name (which again may relate
to different scientists). The usual way to resolve this issue is to refine initial answer
sets with additional searches based on company name or research topic. In relation to
the former, both CAPLUS and MEDLINE list only one affiliation (Section 6.3) so the
refinement by company name may not be comprehensive.

How crucial these issues are depends on the intent of the search and on how unusual the name is. If the intention is to find the majority of papers written by an author, then **Explore: Author Name** followed by refinements is usually sufficient. If the intention is to get a complete list of papers for an author, then all possible variations of the author's name must be considered and it is usually necessary to look through *all* the records to check that papers from different authors are not included.

---

*Comment*

The alternative way to identify authors is to obtain a reference set based on a search using terms in the author's research field, and then **Analyze: Author Name**. Often it helps to display the list in alphabetical order. In this way variations in the name for the author may be identified and this information may then be used to start new searches under **Explore: Author Name**.

---

## 6.3  Explore: Company Name

Searching for company names is a *very* complex task and some issues include:

- Authors may state their company name in a number of ways (e.g. may or may not include names of departments or ZIP codes);
- CAPLUS and MEDLINE may have file specific abbreviations (e.g. Univ/University, Dept/Department, Ltd/Limited);
- Company mergers and acquisitions may lead to changes in company names from time to time;
- Only one affiliation is usually listed in CAPLUS and MEDLINE even though separate affiliations for different authors may be given in the original article.

Information on company names may be found in two ways: either a bibliographic answer set may be obtained (e.g. by **Explore: Research Topic**) and then the answers may either be analysed or refined by company name, or **Explore: Company Name** may be used.

The advantage of **Analyze: Company Name** is that individual listings and variations are indicated in the histogram. The limitations are that the list of options may be long and the initial answer set may not have retrieved all relevant records (particularly because of the policy of listing only a single affiliation). Nevertheless, in this way a reasonable indication of the types of entries present may be obtained and if more general searches through **Explore: Company Name** are required then strategies may be developed based on the terms identified.

On the other hand, both **Refine: Company Name** and **Explore: Company Name** require precise inputs and, as already discussed, this needs to be done with caution because of variations of the entries in the databases.

At present in SciFinder there is no way around the restriction that only one affiliation is listed in the databases. When comprehensive information on the company is required alternative databases that list all affiliations in the original documents may need to be consulted. However, all databases are restricted through the first three points indicated above, and caution needs to be exercised in interpreting the results.

---

*Comment*

**Explore: Company Name** searches the company name field, which also includes the country in which the company is located. CAPLUS and MEDLINE have slightly different policies relating to entry of country names and to the abbreviations used. For example, while records from the United Kingdom are nearly always listed as 'UK' in CAPLUS, almost 90% of records from the United Kingdom are listed as 'UK' in MEDLINE and the remainder are listed as 'U K'. The solution is to look through relevant records to determine the terms used and to construct revised searches accordingly.

---

**Explore: Company Name** is used when the search needs to be started through the name of an organization (Figure 6.3). In the search, SciFinder uses a number of algorithms to identify possible candidates, including searches on the complete company name, on individual words in the company name, and on SciFinder's internal company synonym dictionary. This dictionary of company synonyms means that alternative names for the company may be searched in addition to the names entered.

Generally it is preferable to enter as few terms as possible and to choose terms that may have few synonyms. For example, if records from the Rega Institute Medical Research in Belgium are required then it is advisable to start with **Explore: Company Name** 'Rega' and to work through the initial answers.

Names of institutions may be lengthy, so the full name of the institution may not be shown in the histogram. Additionally, the search field may contain subfields with information on the Department, the Institution, and the Country, and depending on the

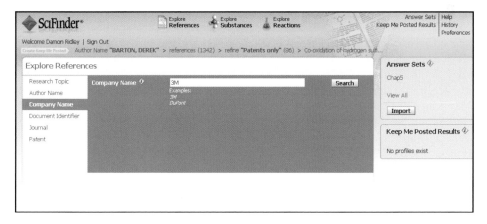

***Figure 6.3*** ***Explore: Company Name*** *screen. Initial search should be general; refinements should be taken in steps. SciFinder® screens are reproduced with permission of Chemical Abstracts Service (CAS), a division of the American Chemical Society*

search different sections may be displayed in the histogram. Accordingly, in practice it is necessary to look through the entire list and then to work systematically through subsequent sets of answers.

If two or more terms are entered in the query, SciFinder looks for these terms anywhere in the company name field and, depending on the specific entries in the company dictionary, may add synonyms to the search. For example, if '3M' is entered then a number of synonyms including 'Minnesota Mining and Manufacturing' will be searched since these are in the company synonym dictionary. However, if '3M Japan' is entered, the algorithm may retrieve records with the terms entered only and synonyms may not be applied. For this reason it is usually better to take the search in steps. For example, a search on '3M' (where the synonym dictionary is applied) followed by a refine with 'Japan' gives almost double the number of answers (218) compared to an initial search just on '3M Japan' (125 answers).

Word order is not important; searches for 'univ sydney', 'university sydney', 'sydney university', and 'sydney univ' produce identical answer sets. However, these records will contain hits for several of the universities in Sydney, including Macquarie University, University of New South Wales, The University of Sydney, and University of Western Sydney. (Hits for the first two universities occur because 'Sydney' is often mentioned in the street address.)

In summary, searching for a company name presents many challenges for the searcher and for the algorithms used by SciFinder. While the algorithms work well in most cases, if comprehensive and precise answers are required, it is probably better to refer the problem to an information professional.

## 6.4   Explore: Document Identifier

Accession numbers are used by database producers as unique identifiers for records. If the searcher wishes to retrieve a record, e.g. to display the full record, to check CAS Registry Numbers, or to work through citation links or full text, then the accession number may be entered directly - if it is known from a previous search.

Patent numbers are used by patent authorities as unique identifiers for patents, but many numbers may relate to the single invention as it passes through the patenting process. For example, a *patent application number* is assigned when the manuscript is first received and a *patent number* is assigned when the manuscript has passed the full examination process. Patents need to be taken out in all countries where the inventor seeks coverage, so a single invention may have many application and patent numbers. All of these numbers make up what is known as the patent (family) information (Figure 6.4) and searches on any of the numbers will retrieve the record.

In this case the original patent was published in English (see the last entry on the right-hand column in Figure 6.4), but the patent family includes foreign language equivalents.

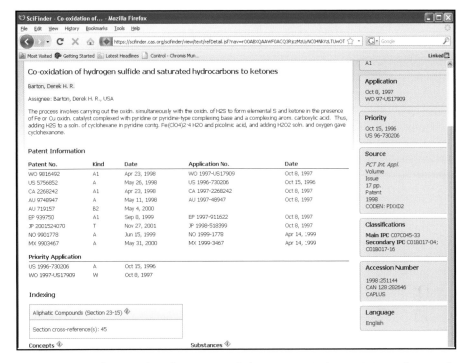

**Figure 6.4** *Part of a record (Reference Detail) for a patent in CAPLUS. Note that members of the patent family are included in the single record. While this basic patent is published in English, some of the other patent family members are published in other languages. SciFinder® screens are reproduced with permission of Chemical Abstracts Service (CAS), a division of the American Chemical Society*

---

*SciFinder Note*

CAPLUS records for patents are usually based on the first released patent/patent application, and the language of that document is recorded in the Language Field. This language will be searched through **Analysis/Refine: Language** or through the additional options located on the initial **Explore References: Research Topic** screen.

Other patent family members may be published in different languages, but currently these additional languages are not searchable in SciFinder. The only way to find whether an original Japanese language patent has an English equivalent is to look through actual records.

---

Scientists may need to consult information professionals for a full explanation of the various patent numbers and codes, although users of SciFinder are generally interested only in the scientific aspects of the document. Indeed, questions related to the legal status of patents most likely need to be asked only if the searcher is interested in the patentability of their research, and such questions are not within the usage restrictions for academic users of SciFinder.

Information on the patents covered in CAPLUS is available through links given in Appendix 1 and SciFinder currently contains records for over 6.5 million patents (i.e. *patent families*). The majority of these records are linked to full text documents. Records have titles, abstracts (note that titles and abstracts in records for patents may be rewritten by the document analysts when the original document does not contain sufficient technical information in these fields), and full indexing, and citations in patents occur from around 1997 onwards, so SciFinder provides many entries into the patent literature at the desktop.

## 6.5   Explore: Journal and Explore: Patent

Records for journals and patents may be searched directly; the **Explore** screens are shown in Figures 6.5 and 6.6 respectively. Entries under **Explore: Journal** may be the full journal name, its abbreviation, or its acronym, although in the last two cases entries must not include spaces or punctuation.

Searches for journals are performed in both bibliographic databases (CAPLUS and MEDLINE); a number of SciFinder algorithms are applied to help overcome the different ways in which journal titles are entered. Nevertheless, it always pays to check answers carefully to ensure that the intended outcomes are obtained.

**Explore: Journal** may be searched periodically in order to view records for current issues. Alternatively, after an initial search, **Create Keep Me Posted** may be used and updates will be sent automatically. In this way titles and abstracts of articles in the latest issues may be browsed conveniently.

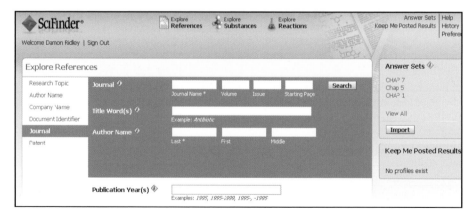

**Figure 6.5   Explore: Journal** *screen. SciFinder*® *screens are reproduced with permission of Chemical Abstracts Service (CAS), a division of the American Chemical Society*

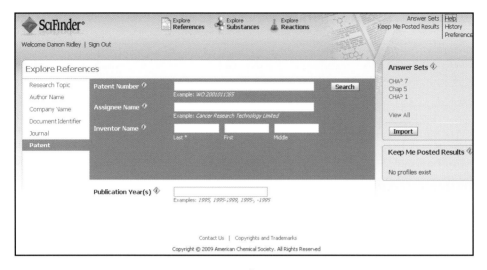

***Figure 6.6　Explore: Patent*** *screen. SciFinder*® *screens are reproduced with permission of Chemical Abstracts Service (CAS), a division of the American Chemical Society*

## 6.6　Getting Information from Bibliographic Records

Screens showing reference answer sets have a number of links to additional information. The options from the references screen are **Get Substances, Get Reactions, Get Cited**, and **Get Citing** (see locator ❸ in Figure 3.3); additionally, **Get Full Text** appears on the reference detail screen (Figure 3.9) and also at the end of each record in the reference summary screen (Figure 3.3).

These options offer ways to *broaden* answers in innovative ways, and a summary of these options is given in Chapter 1, Section 1.2.7. More details are given below.

### 6.6.1　Get Substances

The principle here is that indexed substances (i.e. the CAS Registry Numbers in the bibliographic records) are retrieved in REGISTRY and the substance summary screen (e.g. as shown in Figure 4.4) is displayed. The process proceeds through an intermediary step (e.g. see Figure 3.12), which allows only references that report more specific properties.

In practice, when 'All References' is chosen the process is straightforward. The only issue is the listing of CAS Registry Numbers in the bibliographic databases, and an overview of the policies for registration of substances is given in Section 2.1.2 (for CAPLUS) and in Section 2.2 (for MEDLINE). However, when any of the options under 'References associated with' is chosen, SciFinder additionally checks various bibliographic terms associated with the CAS Registry Number, including CAS Roles in CAPLUS or Allowable Qualifiers in MEDLINE.

**Table 6.1**  *Number of records retrieved when some refinement options are chosen within **Get References** for dihydrotestosterone*

| Option chosen: Get References, then .... | Number of records: CAPLUS | Number of records: MEDLINE |
|---|---|---|
| All references | 9873 | 7324 |
| Adverse effect, including Toxicity | 248 | 236 |
| Analytical study | 436 | 3049 |
| Crystal structure | 43 | 0 |
| Preparation | 130 | 1243 |
| Reactant | 339 | 0 |
| Spectral properties | 150 | 0 |
| Uses | 393 | 1878 |

Subtle issues now come into play, which are best illustrated through an example. The data in Table 6.1 show the number of records in CAPLUS and in MEDLINE when specific references are chosen for dihydrotestosterone (**1**).

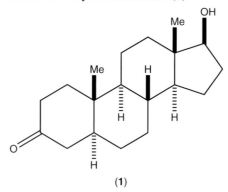

(**1**)

While these data are obtained when **Get References** from a substance record is requested and while **Get Substances** from a bibliographic answer set involves the reverse process, nevertheless the principle is similar and the outcomes reflect the impact of indexing. For example, in this case there are almost 10 times the number of records retrieved for 'Preparation' in MEDLINE than there are in CAPLUS; some of the MEDLINE records will relate to biosynthetic pathways, while some of the CAPLUS records will relate to organic syntheses. However, the other issue is that the CAS Registry Number for dihydrotestosterone is used in MEDLINE to cover a *class of substances* whereas indexing of 'dihydrotestosterones' in CAPLUS is done at the specific substance level. (Since there may be several different compounds in the class defined by MEDLINE, then many additional preparations may be retrieved.)

### 6.6.1.1   Using Get Substances

Chapters 4 and 5 focused on the process where *substances were found in REGISTRY first* and then information on the substances was found in CAPLUS and MEDLINE. Often the specific information required involved chemical or physical properties. The process, by which *bibliographic records are obtained first* and then **Get Substances** is chosen, works in exactly the opposite way. It enables substances with specific properties to be found.

The search may start with a great variety of 'properties'. For example, the property may be:

---

- IR-absorbing compounds;
- Chinese traditional medicines;
- shape-memory compounds;
- jet-lag (medications);
- anti-AIDS agents;

- Grubbs' catalysts;
- spectral information;
- beta-blockers;
- rattlesnake toxins;
- ... almost anything!

---

The process is to find all records through the appropriate **Explore References: Research Topic** and then to click **Get Substances**. A limitation is that fewer than 1000 references may be processed through this function at one time so it may be necessary to narrow initial answers first to come within system limits.

Of course, not all of the substances retrieved will have the required property. For example, **Explore References: Research Topic** 'ginseng with steroid' retrieves around 240 records where the terms are 'anywhere in the reference'. Then, **Get Substances** gives over 6000 substances, including a large number of proteins and nucleic acids that have been found in ginseng and other substances such as calcium ion, water, and carbon dioxide that have been involved in the studies. The solution is to use the **Analyze/Refine** tools to narrow substances, and in this example the first step would be to **Refine: Chemical Structure** where the structure query contains the steroid ring system.

Since a specific type of ring system is of interest, the question is 'why wouldn't the user start with a substructure search, obtain substances, then **Get References**, and finally **Refine: Research Topic** "ginseng"?' This approach could indeed be used, provided system limits are not exceeded; it is just that starting with the property may be easier in this case.

However, there are other cases where the property (e.g. IR-absorbing compounds) is the key requirement and substances of *any structure* are of interest. Now the search needs to start with the property and follow the process described above. This option fully integrates all the unique functionalities for processing answers in CAPLUS/MEDLINE with those in REGISTRY.

### 6.6.2 Citations

Citations have been added to CAPLUS from around 1998, but there are a number of citations in earlier records. While there are a number of issues relating to citations (e.g. see 'Citation searches in on-line databases: possibilities and pitfalls' in *Trends in Analytical Chemistry*, 2001, **20**, 1–10), nevertheless citation linking and citation searching provide further opportunities for information retrieval.

Since answer sets that involve cited or citing references can quickly become very large, the ability to work through larger answer sets in systematic ways with the many SciFinder post-processing tools is important.

```
┌──────────────────────────────────────────────────────────────────────┐
│ Citations                                                              │
│                                                                        │
│ 1) Paolesse, R; The Porphyrin Handbook 2000, V2, P201                  │
│ 2) Erben, C; The Porphyrin Handbook 2000, V2, P233                     │
│ 3) Adamian, V; Inorg Chem 1995, V34, P532                              │
│ 4) Gross, Z; Chem Commun 1999, P599                                    │
│ 5) Simkhovich, L; Inorg Chem 2000, V39, P2704                          │
│ 6) Tardieux, C; J Heterocycl Chem 1998, V35, P965                      │
│ 7) Jerome, F; Chem Commun 1998, P2007                                  │
│ 8) Jerome, F; New J Chem 1998, V22, P1327                              │
│ 9) Paolesse, R; Inorg Chem 1994, V33, P1171                            │
│ 10) Kadish, K; J Am Chem Soc 1998, V120, P11986                        │
│ 11) Guilard, R; Inorg Chem 2001, V40, Pxxxx                            │
│ 12) Lin, X; Anal Chem 1935, V57, P1498                                 │
│ 13) Ellis, J; J Am Chem Soc 1980, V102, P1889                          │
│ 14) Rougee, M; Biochemistry 1975, V14, P4100                           │
│ 15) Barton, D; Tetrahedron 1990, V46, P7587                            │
│ 16) Conlon, M; J Chem Soc, Perkin Trans 1 1973, P2281                  │
│ 17) Hitchcock, P; J Chem Soc, Dalton Trans 1976, P1927                 │
│ 18) Nonius, B; OpenMoleN, Interactive Structure Solution 1997          │
│ 19) Friedman, M; J Org Chem 1965, V30, P859                            │
│ 20) Dumoulin, H; J Heterocycl Chem 1997, V34, P13                      │
│ 21) Paolesse, R; Inorg Chim Acta 1993, V203, P107                      │
│ 22) Scheidt, W; J Am Chem Soc 1973, V95, P8289                         │
│ 23) Boschi, T; J Chem Soc, Dalton Trans 1990, P463                     │
│ 24) Liccocia, S; Inorg Chem 1997, V36, P1564                           │
│ 25) Adamian, V; Ph D Dissertation, University of Houston 1995          │
│ 26a) Geiger, W; Advances in Organometallic Chemistry 1985, V24, P89    │
│ 26b) Kotz, J; Topics in Organic Electrochemistry 1986, P95             │
│ 27) Yap, W; J Electroanal Chem Interfacial Electrochem 1981, V130, P3  │
│ 28) Kadish, K; Inorg Chem 1994, V33, P471                              │
│ 29) Chang, D; Inorg Chem 1984, V23, P1629                              │
│ 30) Kadish, K; Manuscript in preparation                               │
│ 31) Murakami, Y; Bull Chem Soc Jpn 1978, V51, P123                     │
│ 32) Herlinger, A; J Am Chem Soc 1971, V93, P1790                       │
│ 33) Mu, X; Inorg Chem 1989, V28, P3743                                 │
│ 34) Hu, Y; Inorg Chem 1991, V30, P2444                                 │
│ 35) Schmidt, E; J Am Chem Soc 1996, V118, P2954                        │
└──────────────────────────────────────────────────────────────────────┘
```

**Figure 6.7**  *List of citations in CAPLUS for the record AN 2001:643612. There are a total of 35 citations and links are available to records in SciFinder for 26 of these. SciFinder*® *screens are reproduced with permission of Chemical Abstracts Service (CAS), a division of the American Chemical Society*

### 6.6.2.1   Cited References (Working Backwards in Time)

Cited references are displayed automatically as part of the record. Figure 6.7 shows citations for the record AN 2001:643612. Links to records in the database are provided for each citation and the absence of links indicates a record for the citation in the original document is (possibly) not present. The uncertainty (i.e. 'possibly') is due to the fact that these links are obtained through application of computer algorithms that match the citation in the article to records in the database and citations may not be always identified (e.g. incorrect volume or page numbers are given in the original document, or for books the page number given in the original document may not be for the first page of the book or of the chapter).

While citations may be viewed in turn through the links, **Get Cited** gives the complete list of cited records as a separate answer set. This answer set may be analysed or refined in the usual manner, which is important since at times some citations may refer not to the subject matter (which usually is the intention in finding related records) but to other aspects (such as experimental details, melting points, and so forth).

For example, **Explore References: Research Topic** 'urocanic acid with cancer' gives just over 150 records where the two concepts are 'anywhere in the reference', while **Get**

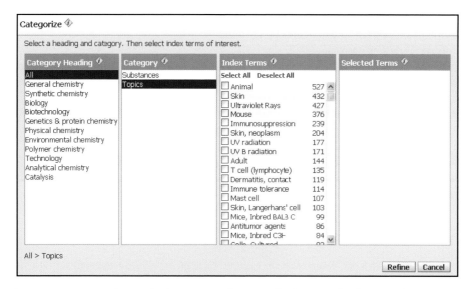

***Figure 6.8*** *Screen obtained when **Categorize** is applied to 'cited references' from answers through **Explore References: Research Topic** 'urocanic acid with cancer'. SciFinder® screens are reproduced with permission of Chemical Abstracts Service (CAS), a division of the American Chemical Society*

**Cited** gives over 1900 records. Cited references are linked to database records, so this answer set will contain records from CAPLUS and MEDLINE.

Advantage may now be taken of the **Analyze/Refine/Categorize** options; in Figure 6.8 an example is shown of the output when **Categorize** is chosen for these cited references. Relevant terms from the list of Index Headings shown are then chosen; clearly many of these would be of interest for an original search based on urocanic acid and cancer.

### 6.6.2.2 Citing References (Working Forward in Time)

Working with citing references enables a searcher to investigate more recent information on the topic. As citations are presently included only in CAPLUS, citing references give CAPLUS records only. Thus after the records that contain the concepts 'urocanic acid' and 'cancer' are retrieved and **Get Citing** is chosen, a new answer set of around 790 records is obtained and the majority of these give more recent information on the subject. (The output, when **Categorize** is chosen for these citing references, is similar to that shown in Figure 6.8.)

Expanding searches through citations is another way of working through issues of search comprehension and precision. It is simply another tool that is available, but it relies ultimately on the references used by the authors in the original publications. At times it turns up important additional articles that may not have been retrieved even by a variety of searches on the topic.

Another application of **Get Citing** (references) is to locate citations to works by individual authors. First all records published by the author are found (Section 6.2) and after **Get Citing** is clicked the required citing references are presented. Since at present a maximum of 500 answers can be processed at one time, it may be necessary to break the initial answer set of publications into subsets, which may easily be done through **Analyze: Publication Year**.

As with all searches it is necessary to understand the limitations, and here the main considerations are the accuracy with which the original search is conducted (e.g. note that it may be difficult to get all the publications by the author and at the same time not to have false answers because of problems with different authors being represented by the same name terms), the years for which citations are entered in the database, and the identification of a record in the database for the citing reference. There are many issues relating to this last aspect including the accuracy of the original citation and the function of the algorithm, so the number of records obtained through the **Get Citing** function should be considered as a guide only.

## 6.7    Further Issues with Finding Information on Substances

In Chapters 4 and 5 the focus was on finding substances initially through REGISTRY (i.e. by name, formula, or structure search terms), but substances may also be found through **Explore References: Research Topic**. For example, if the user is interested in information on the antibiotic/immunosuppressant cyclosporin A and cancer, alternative approaches may be:

(a) **Explore References: Research Topic** 'cyclosporin A with cancer'

*or*

(b) **Explore Substances**, then **Substance Identifier** 'cyclosporin A', then **Get References** followed by **Refine References** with 'cancer'.

The preferred option depends on the actual search and often both options may need to be investigated. As is often the case, it is best to do the experiment!

### 6.7.1    Option (a). Starting with *Explore References: Research Topic*

In this case, when option (a) is chosen, SciFinder indicates there are over 650 records in which the concepts 'cyclosporin A' and 'cancer' are 'closely associated' and over 3700 records in which the concepts are 'anywhere in the reference'. A quick check on some of the answers indicates that synonyms for cancer such as neoplasm and carcinoma are searched automatically, while terms retrieved for cyclosporin A in CAPLUS and in MEDLINE included the CAS Registry Number for cyclosporin A (59865-13-3) as a hit term. So all looks good!

However, it is noted that the MEDLINE Index Heading 'Cyclosporine' is not a hit term (i.e. it is not highlighted in MEDLINE records that are retrieved). This may be surprising since it may have been expected that application of the truncation algorithm in **Explore References: Research Topic** would have retrieved 'cyclosporine'. Then, on further reflection, the searcher may wonder why the concept 'cyclosporin A' was identified anyway since the letter 'A' is a stop-word (i.e. a word, like a preposition, that SciFinder ignores when concepts are being identified).

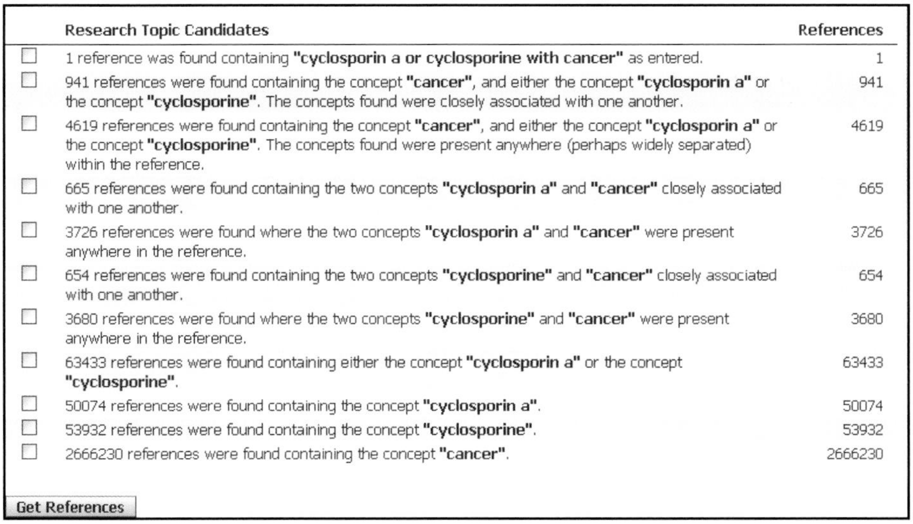

| Research Topic Candidates | References |
|---|---|
| ☐ 1 reference was found containing **"cyclosporin a or cyclosporine with cancer"** as entered. | 1 |
| ☐ 941 references were found containing the concept **"cancer"**, and either the concept **"cyclosporin a"** or the concept **"cyclosporine"**. The concepts found were closely associated with one another. | 941 |
| ☐ 4619 references were found containing the concept **"cancer"**, and either the concept **"cyclosporin a"** or the concept **"cyclosporine"**. The concepts found were present anywhere (perhaps widely separated) within the reference. | 4619 |
| ☐ 665 references were found containing the two concepts **"cyclosporin a"** and **"cancer"** closely associated with one another. | 665 |
| ☐ 3726 references were found where the two concepts **"cyclosporin a"** and **"cancer"** were present anywhere in the reference. | 3726 |
| ☐ 654 references were found containing the two concepts **"cyclosporine"** and **"cancer"** closely associated with one another. | 654 |
| ☐ 3680 references were found where the two concepts **"cyclosporine"** and **"cancer"** were present anywhere in the reference. | 3680 |
| ☐ 63433 references were found containing either the concept **"cyclosporin a"** or the concept **"cyclosporine"**. | 63433 |
| ☐ 50074 references were found containing the concept **"cyclosporin a"**. | 50074 |
| ☐ 53932 references were found containing the concept **"cyclosporine"**. | 53932 |
| ☐ 2666230 references were found containing the concept **"cancer"**. | 2666230 |

Get References

*Figure 6.9*  *Candidate screen from **Explore References: Research Topic** 'cyclosporin a or cyclosporine with cancer'. The different numbers of references for the single concepts 'cyclosporin a' and 'cyclosporine' indicate the concepts are different. SciFinder® screens are reproduced with permission of Chemical Abstracts Service (CAS), a division of the American Chemical Society*

Somewhat worried, the user may then perform a couple of other experiments: first, **Explore References: Research Topic** 'cyclosporin A or cyclosporine with cancer' and, second, **Explore References: Research Topic** 'cyclosporin or cyclosporine with cancer', since these may help to understand what is occurring. The Research Topic Candidates for these two experiments are shown in Figures 6.9 and 6.10 respectively.

While there are several points to note, perhaps the most critical is that the concepts 'cyclosporin a', 'cyclosporine', and 'cyclosporin' have similar, *but different*, numbers of references. (If **Explore References: Research Topic** 'cancer or neoplasm or carcinoma' is undertaken, then it is found that the number of references for each of the concepts is identical, 2,666,230.) Thus it is apparent that the different 'cyclosporin' terms do not make up part of the same concept, and the question is 'why?'

The broad answer is that the developers of SciFinder always try to balance comprehension with precision. To do this, outcomes of the different **Explore References: Research Topic** algorithms are analysed, and those which on average give the best balance are employed. A key issue is what to do when a 'concept' (identified as explained in Section 3.3) matches exactly either the name of a specific substance in REGISTRY or an Index Heading in CAPLUS or in MEDLINE. When this occurs, the algorithm is usually stopped from proceeding further, e.g. to the application of automatic truncation, since it is felt that once an index entry (CAS Registry Number or Index Heading) is identified then these offer a good compromise between comprehension and precision anyway.

| Research Topic Candidates | References |
|---|---|
| ☐ 1 reference was found containing **"cyclosporin or cyclosporine with cancer"** as entered. | 1 |
| ☐ 1027 references were found containing the concept **"cancer"**, and either the concept **"cyclosporin"** or the concept **"cyclosporine"**. The concepts found were closely associated with one another. | 1027 |
| ☐ 5045 references were found containing the concept **"cancer"**, and either the concept **"cyclosporin"** or the concept **"cyclosporine"**. The concepts found were present anywhere (perhaps widely separated) within the reference. | 5045 |
| ☐ 752 references were found containing the two concepts **"cyclosporin"** and **"cancer"** closely associated with one another. | 752 |
| ☐ 4163 references were found where the two concepts **"cyclosporin"** and **"cancer"** were present anywhere in the reference. | 4163 |
| ☐ 654 references were found containing the two concepts **"cyclosporine"** and **"cancer"** closely associated with one another. | 654 |
| ☐ 3680 references were found where the two concepts **"cyclosporine"** and **"cancer"** were present anywhere in the reference. | 3680 |
| ☐ 67979 references were found containing either the concept **"cyclosporin"** or the concept **"cyclosporine"**. | 67979 |
| ☐ 54838 references were found containing the concept **"cyclosporin"**. | 54838 |
| ☐ 53932 references were found containing the concept **"cyclosporine"**. | 53932 |
| ☐ 2666230 references were found containing the concept **"cancer"**. | 2666230 |

Get References

**Figure 6.10**  *Candidate screen from* **Explore References: Research Topic** *'cyclosporin or cyclosporine with cancer'. The different numbers of references for the single concepts 'cyclosporin' and 'cyclosporine' indicate the concepts are different. SciFinder® screens are reproduced with permission of Chemical Abstracts Service (CAS), a division of the American Chemical Society*

---

*Comment*

It may be good if SciFinder always gave the 'perfect' interpretation of the user's query; the user then would never need to know what occurred, or why, and would never have to 'learn' about information retrieval. That may be convenient, even if it may not satisfy scientific curiosity.

However, computer algorithms work 'perfectly' only when consistently written data always needs to be processed in the same way. Computers are great at number crunching! The reality is that text written by authors in titles and abstracts, and the interpretation of the science and indexing by document analysts, may not be 'perfect'.

Furthermore, the type of information that a searcher thinks is needed to solve a problem (and hence the way the initial question is phrased) may not always be 'perfect'; indeed experienced searchers know that they may well end an online search session with quite a different outcome from what may have been conceived at the beginning.

Accordingly, the philosophy of SciFinder is to interpret the initial query in a consistent way and then to give the user the tools to narrow, or broaden, the answers in ways that are well understood.

In the example just given ('cyclosporin A with cancer') SciFinder has made a great start. The user then needs to apply the basic principles of scientific method – particularly to check for indexing and to check the SciFinder interpretation of the query – and finally to use SciFinder tools efficiently and creatively.

From the information displayed in Figures 6.9 and 6.10, and from the knowledge gained by observing some answers that concepts 'cyclosporin a' and 'cancer' include the CAS Registry Number 59865-13-3 and synonyms for 'cancer', it may be decided that the most appropriate **Explore References: Research Topic** may be 'cyclosporin a or cyclosporin or cyclosporine with cancer'.

The next question is whether to choose the candidates 'closely associated' or 'anywhere in the reference'. As noted previously, separate Index Headings in a record are *not* considered to be 'closely associated', so particularly if MEDLINE is considered to be the preferred database, then candidates 'anywhere in the reference' probably need to be chosen. The same applies for Index Headings in CAPLUS, although because of the inclusion of text-modifying phrases with Index Headings the candidates 'closely associated' may offer more precise answers without sacrificing comprehension greatly.

### 6.7.2   Option (b). Starting with *Explore Substances*

When option (b) is chosen, the substance cyclosporin A is easily found (e.g. through the name under **Substance Identifier**) and **Get References** affords over 40,000 records. When these are refined with 'cancer' there are almost 3000 answers, of which about one-half are in each of CAPLUS and MEDLINE.

### 6.7.3   Further Considerations

A further option now worth consideration is **Explore References: Research Topic** '59865-13-3 with cancer' since this will give a candidate answer set in which the two terms are 'closely associated'. In fact there are almost 200 records for this candidate and of course all are from CAPLUS (since CAS Registry Numbers in MEDLINE are in a separate field/sentence).

So perhaps **Explore References: Research Topic** 'cyclosporin A (59865-13-3) with cancer' and then selection of the 2000+ records in which the different concepts are 'anywhere in the reference' may be a useful starting point. From there, **Analyze/Refine** options may be used to select more specific answers.

While this example outlines some searches for the topic 'cyclosporin A with cancer', naturally searches under options (a) or (b) for different substances/topics will turn up other variations. The key issues for the searcher are:

• The impacts of searching for CAS Registry Numbers in the different databases;
• How the algorithm searches under **Explore References: Research Topic** for the various 'words' in the name for the substance (e.g. **Explore References: Research Topic** '4-nitropyridine' will identify concepts '4' and 'nitropyridine', and even candidates where they are 'closely associated' may contain the concepts in different parts of the sentence);
• How those 'words' appear in actual records (e.g. searches on the words 'benzoic acid' will retrieve records with entries such as 'benzoic acid esters);
• Whether to search for concepts 'closely associated' or 'anywhere in the reference'.

In summary, searching for substances *by name* in **Explore References: Research Topic** may lack precision but may aid comprehension. On the other hand, finding substances by name starting from **Explore Substances** will always employ CAS Registry Numbers in subsequent steps. These are precise search terms, which also aid in comprehension since they may effectively cover many names for a substance, and may be the only entry point for a substance in a record (e.g. the CAS Registry Number 9068-38-6 in Figure 3.9).

## 6.8    Opportunities for MEDLINE Searchers

The MEDLINE database may be searched through many sources, some of which are listed in Appendix 1, and so the ability to search MEDLINE through SciFinder offers another interface that the medical researcher needs to evaluate. While individual users need to make these evaluations for their own research, some general comments on the implementation through SciFinder follow.

### 6.8.1    Complimentarity of MEDLINE and CAPLUS

SciFinder users may search MEDLINE and CAPLUS simultaneously and thus take full advantage of their complementarity.

#### 6.8.1.1    Unique Records

Only following a most detailed analysis of records may a full understanding of the unique coverage in each database be achieved, but in general MEDLINE may offer more comprehensive coverage in the areas of clinical, social, and epidemiological medicine. Both databases cover pre-clinical medicine, but as the studies become more molecular then CAPLUS increasingly becomes the more important resource. Indeed, the Biological Section Codes of CAPLUS contain more than 9.7 million records (almost one-third of the database) covering pre-clinical research, including biochemistry, pharmacology, and molecular biology.

   CAPLUS also compliments MEDLINE in that CAPLUS contains over 560,000 records (patent families) in the Biological Sections from the patent literature. The number of patents in the Biological Sections in CAPLUS is growing rapidly and over one-third of these patent records have been added since 2006.

---

*Comment*

An advantage of searching CAPLUS for patent information in the biomedical sciences is that CAPLUS records for patents have very substantial subject and substance/ sequence indexing. Currently more substances from patents are entered in REGISTRY each year than substances from journals or any other source; a majority of these new substances relate to the biomedical sciences.

   Note also that finding substances precisely and comprehensively in full text sources is challenging. In the case of patents many different substance names may be used and additionally full text documents may mention many substances for which new information is not reported.

---

### 6.8.1.2   Unique Indexing

Where there is overlap in coverage, the indexing is quite different (e.g. Figures 1.5 and 1.7) and the ability to search two sets of index terms may substantially increase recall and precision. Additionally, identification of index terms may often alert the user to terms that may be added to the initial question. For example, **Explore References: Research Topic** 'inhibition of hiv replication in humans', followed by **Analyze: Index Term**, indicates important headings are 'antiviral agents', 'viral replication (drug effects)', and 'HIV-1 (drug effects)'. These are MEDLINE Index Headings and would not have been retrieved through the initial search, so it would now be advisable to revise the search to include these terms.

### 6.8.1.3   Display and Duplicate Records

By default answers are presented in Accession Number order with CAPLUS records followed by MEDLINE records, although there are other Sort options (Author Name, Publication Year, and Title); the use of these was mentioned earlier (Section 3.5.3).

If answers from only one database are required then **Analyze: Database** gives the necessary choices. Duplicate records may be removed with the function **Remove Duplicates**, but this is best done after all analysis and refinement options, particularly those relating to indexing, have been explored.

### 6.8.1.4   CA Lexicon and MeSH Thesaurus

The identification of Index Headings is achieved mainly at the secondary level in SciFinder; i.e. a bibliographic answer set is obtained first and the Index Headings that are identified may be used either directly to narrow the search or as a guide to terms to use in a revised initial search. In fact, this approach serves the user very well, since it allows the searcher to take advantage of indexing without having to know too many details of index policies in advance.

However, both CAPLUS and MEDLINE have Index Headings arranged in hierarchical structures (CA Lexicon, Section 2.1.2, and MeSH thesaurus, Section 2.2) and examples are given in Figures 2.3 and 2.6 respectively. The advantage of these hierarchical index structures is that searches at broader, narrower, or associated index levels may be conducted, and these may facilitate the retrieval of comprehensive and precise answer sets. Accordingly, it helps if users keep in mind that there are a variety of information retrieval strategies, but the use of thesaurus capabilities is a specialized topic and should be explored with help from information professionals.

## 6.8.2   Complimentarity of REGISTRY, MEDLINE, and CAPLUS

Currently REGISTRY has more than 6.7 million protein sequences and more than 55 million nucleic acid sequences. There are also more than 57,000 different CAS Registry Numbers in MEDLINE, which appear in more than 5.7 million MEDLINE records. Meanwhile, CAPLUS has more than 7.7 million records that list the CAS Role: Biological Study, and the numbers of records with the specific Roles: Adverse Effect and Therapeutic Use are more than 700,000 and more than 1 million respectively. These data give an idea of some of the relationships between the three major databases in SciFinder.

---

*Statistics!*

What do these numbers mean for biomedical searchers?

CAS Registry Numbers provide precise search terms for substances/sequences and they may be found in several ways through SciFinder. Approximately one-half of MEDLINE records have CAS Registry Numbers.

Approximately one-quarter of the records in CAPLUS are indexed with the specific CAS Role Biological Study, and many other CAS Roles relate to biological properties. In turn most of these CAS Roles are associated with CAS Registry Numbers.

The links between REGISTRY and CAPLUS/MEDLINE in SciFinder offer enhanced search capabilities!

---

Substances may be retrieved in SciFinder through name, formula, or structure searches, and the latter in particular present opportunities that are not readily available to MEDLINE searchers. As an example of the types of opportunities in the medical sciences presented through structure searches, consider the situation where a scientist undertook **Explore References: Research Topic** 'the inhibition of gram positive bacteria' and found, among other things, a patent (WO9953915) that describes the use of the furanone (**2**). The question is 'What other furanones have activity against gram positive bacteria?'

(2)

In the event, when a substructure search was performed on the parent furanone query (Figure 6.11) and when 'Conventional Substructures' was chosen, an answer set of almost 5000 substances was retrieved. However, users should not necessarily be put off by large answer sets, since SciFinder has excellent ways to analyse/refine answers. In this case, while the substances may be narrowed with the help of the various analysis tools discussed in Section 5.3, the ultimate intention is to restrict *references* to those that mention gram positive bacteria, and at this final stage probably relatively few references will result. So **Get References** was chosen and then **Refine: Research Topic** 'gram positive' gave an answer set of 37 records, of which 31 were from CAPLUS and six from MEDLINE. These records contained just over 100 substances with furanone structures (some are shown in Figure 6.12) and as many of these are commercially available the biomedical scientist would readily be able to obtain a number of furanones for testing.

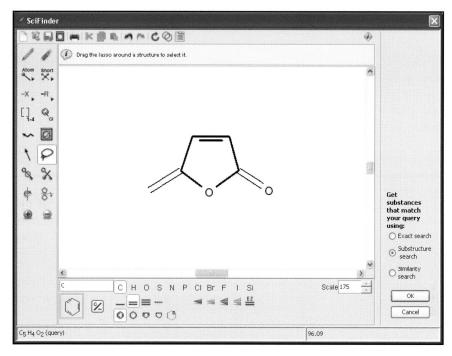

***Figure 6.11*** *Structure query to start a search for references that looks for furanones with activity against gram positive bacteria. SciFinder® screens are reproduced with permission of Chemical Abstracts Service (CAS), a division of the American Chemical Society*

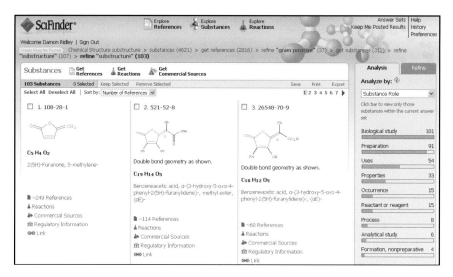

***Figure 6.12*** *Some structures of furanones that have activity against gram positive bacteria. SciFinder® screens are reproduced with permission of Chemical Abstracts Service (CAS), a division of the American Chemical Society*

Even though this is an excellent result, it helps if the user considers the possible limitations of the search and considers what alternative strategies may be used. In this case, a precise search on a part-structure was undertaken, which would have comprehensively and precisely covered the 'furanone' concept in the original query. Of course, the answers would be restricted to those records that listed the CAS Registry Numbers, but it is reasonable to assume that, if something new about the antibiotic property of furanone derivatives was mentioned in the original article, then the CAS Registry Numbers would have been indexed.

However, the refinement 'gram positive' may not have been comprehensive, and in particular refinement with the names of specific bacteria may be considered. Once again, clues on how to proceed may be gained through SciFinder, e.g. through analysis of the 37 answers by Index Term. When this is done, the names of a number of specific bacteria become apparent and these may be entered as an alternative refinement for the initial answer set. Indeed, when **Refine: Research Topic** 'gram positive or enterobacter or staphylococcus or streptococcus' was applied to the references from the substances retrieved in the original substructure search, 85 bibliographic records were retrieved. This is an even better result!

This example illustrates the value of the ability to search for substances in REGISTRY in a precise manner, and then to search for information on these substances in both CAPLUS and MEDLINE. Effectively, the process achieved 'a substructure search in MEDLINE'. Links between the CAS substance and bibliographic databases and the MEDLINE database provide medical scientists with opportunities to find more precise and comprehensive information, and in creative ways.

### 6.8.3   The SciFinder Interface and Search Opportunities

The SciFinder interface guides the user and considerably assists with search retrieval in many areas such as topics, authors, companies, and substances. Once the overall philosophy and function of SciFinder is understood, scientists may quickly and effectively accomplish excellent search results.

Further, the manner in which SciFinder integrates the databases provides many search opportunities, particularly through the ability to **Get Substances** from bibliographic answers. Therefore, if the medical scientist is interested in substances to which the malaria parasite is not resistant, then **Explore References: Research Topic** 'malaria parasite or plasmodium falciparum with drug resistant' affords around 1500 records in CAPLUS. These records may be analysed by CAS Registry Number to obtain a histogram of substances, the substances may be retrieved through **Get Substances**, or various options under **Categorize** may be used (one example is shown in Figure 6.13). The substances in column 3 in this figure may not all be active against strains of the parasite, but still this process gets the user started – and hopefully suggests new research areas.

In summary, SciFinder offers many opportunities for those who traditionally have used MEDLINE!

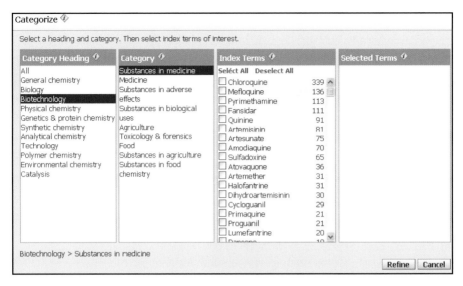

**Figure 6.13**   *Example of screen obtained in **Categorize** for search on drug resistant malaria parasites. SciFinder® screens are reproduced with permission of Chemical Abstracts Service (CAS), a division of the American Chemical Society*

## 6.9   Searching for Substances in the Biological Sciences

While searching for organic and inorganic substances presents a number of challenges, nevertheless very often the scientist is dealing with a discrete substance, and some of the options discussed in Chapters 4 and 5 will usually produce excellent answers. On the other hand, searching for 'biological' substances often presents different challenges because the exact nature of the substance may be unknown, or if it is known then it may be difficult to describe. For example, how are genes, or plasmids, or immunoglobulins described? How may genetic modifications be retrieved? Transport proteins? Antibodies? Receptors? Cloned DNA?

When answering questions of this type it helps if the scientist goes back to basic principles, namely that entries in bibliographic databases are written in part by authors and in part by indexers. The challenge is to anticipate terms used by authors or to know how the field is covered by indexers. Actually neither option is that daunting, since scientists are aware of author terminologies in their area and index terms may be identified through **Analysis: Index Term** or **Categorize** once an initial answer set has been obtained.

Therefore, armed with knowledge of author terminology and knowing that indexing can be found through the use of SciFinder post-processing functions, the best solution is to try some options. Nevertheless, some general aspects of indexing may be helpful and summaries are given below.

### 6.9.1   Nucleic Acids and Related Terms

Table 6.2 summarizes indexing in CAPLUS of some topics in the field of molecular biology. The first two columns present *some* general Index Headings in the field and the approximate time periods for which they were valid, the third column indicates the number of records with the various headings, while the last column indicates *some* alternative terms and notes. Thus the Index Heading 'Deoxyribonucleic acids' was used for general studies on this class of substances between 1972 and 1996, and just over 137,000 records have this heading. Since 1996 the Index Heading 'DNA' has been used and the heading already appears in over 140,000 records. Similarly, the Index Heading 'Ribonucleic acids' and its various subsets have been replaced by 'RNA', 'mRNA', and so forth.

Often within these general headings are many narrower headings that may be even more commonly listed. For example, 'Genetic methods' appears in over 8000 records whereas the narrower headings 'Genetic mapping', 'Genetic vectors', and 'Nucleic acid hybridization' each appear in well over 30,000 records. The indexing principle in CAPLUS is that as *precise* headings as possible are applied, except for very general studies that are indexed at the broader level. Accordingly, it cannot be assumed that searches on general headings alone will cover all the more detailed information on the topic.

It will be apparent from Table 6.2 that some of the changes in indexing will be of little consequence to the user of SciFinder. Thus **Explore References: Research Topic** 'microbial genes' will retrieve 'gene, microbial'. However, will an **Explore** 'dna sequences' include 'deoxyribonucleic acid sequences' or 'cDNA sequences' as search terms? And will an **Explore** 'mRNA' include 'messenger ribonucleic acids' in the search? The answer is simple: 'Just try it!'

One point in favour of searching for nucleic acids and derivatives is that there are very few building blocks (e.g. there are only a few nucleic acid bases) and relatively few functions (e.g. most simply provide the code to manufacture proteins). Further, molecular biologists have developed simple ways to describe 'substances'. For example, the isoleucine valine genes and operons (and related proteins) have letters or numbers to denote specific types such as ilvA, ilvB, ilv1, ilv3, ilvGMEDA, and so forth. Generally these terms are very specific; searches on these terms through **Explore References: Research Topic** and then a choice of candidate 'as entered' or the 'concept' will quickly lead the scientist to precise answers.

### 6.9.2   Proteins

The same cannot be said about proteins, and particularly about enzymes for which numerous different terms are used. In general, proteins are indexed either as the specific sequence or as the protein class (or function).

The general format for the systematic naming of proteins with a specific sequence in REGISTRY is:

**Product (source | strain cell tissue | clone | descriptor | form | other).**

**Table 6.2** *Some major Index Headings in CAPLUS in the field of molecular biology*

| Index Heading | Valid periods | Number of records | Some alternative terms (numbers of records) and notes |
|---|---|---|---|
| Chromosome | 1972– | >100K | Used for: Genetic maps<br>Homologous chromosome |
| Deoxyribonucleic acids | 1972–1996 | >137K | Replaces: Nucleic acids, deoxyribo-<br>New term: DNA (>140K) |
| DNA sequences | 1997– | >125K | Old term: Deoxyribonucleic acid sequences (>90K)<br>Related term: cDNA sequences (>92K) |
| Gene | 1962– | >715K | Used for general studies on genes |
| Gene and genetic element | 1982–1991 | >113K | New term: Gene (>715K)<br>New term: Genetic element (>140K) |
| Gene, animal | 1992– | >435K | Old term: Gene and Genetic element, animal (>57K) |
| Gene, plant | 1992– | >59K | Old term: Gene and Genetic element, plant (>7K) |
| Gene, microbial | 1992– | >190K | Old term: Gene and Genetic element, microbial (>50K) |
| Genetic element | 1987– | >140K | Old term: Gene and genetic element (>114K) |
| Genetic engineering | 1982– | >28K | Narrower term: Molecular cloning (>108K) |
| Genetic methods | 1992– | >8K | Narrower terms include: Genetic mapping (>58K), Genetic vectors (>31K), Nucleic acid hybridization (>46K) |
| Molecular cloning | 1977– | >108K | Used for: Cloning<br>Gene fusion |
| Nucleic acids | 1977– | >61K | The general heading has >40 narrower and related terms |
| Nucleosides | 1907– | >21K | >60 narrower and related terms |
| Nucleotides | 1907– | >44K | >100 narrower and related terms |
| Plasmids | 1967– | >16K | Old term: Plasmid and episome (>36K) |
| Promoter (genetic element) | 1997– | >75K | |
| Ribonucleic acids | 1967–1996 | >130K | New term: RNA (>39K) |
| Ribonucleic acids, messenger | 1972–1996 | >130K | New term: mRNA (>89K) |
| Ribonucleic acids, ribosomal | 1972–1996 | >13K | New term: rRNA (>27K) |
| Ribonucleic acids, transfer | 1972–1996 | >11K | New term: tRNA (>11K) |
| Ribonucleic acids, viral | 1972–1996 | >11K | New term: viral RNA (>11k) |
| Transcription factors | 1997– | >190K | >30 narrower and related terms |
| Transcription, genetic | 1992– | >59K | Used for: Transcription<br>Genetic transcription |
| Transcriptional regulation | 1997– | >77K | |

So the product (transcription factor Tfam) from the source (Rattus norvegicus), strain cell tissue (strain Sprague-Dawley), and clone (clone 1) with the descriptor (precursor) has the systematic name (see Appendix 4, Section A4.4.5):

**Transcription   factor Tfam (Rattus norvegicus strain Sprague-Dawley clone 1 precursor)**.

There are exceptions to this pattern for immune proteins, fragment proteins, fusion proteins, and chemically modified proteins, but the alternate indexing may be understood by looking through some relevant records.

Many protein classes such as enzymes, toxins, hormones, most growth factors, antimicrobial peptides, cytochromes, and hemoglobins are indexed as Index Headings that relate to their function. Indeed, the general rule is that proteins that exhibit clear 'activity' receive functional registration at the Index Heading level.

However, there are many types of proteins that do not receive functional registrations. Examples include receptors, viruses, transport proteins, calcium-binding proteins, and signalling intermediaries. Thus, receptors are indexed through the chemicals or ligands to which they respond (e.g. aspartate receptors, frizzled receptors, toll receptors), and while 'Transport proteins' is an Index Heading in CAPLUS (with over 58,000 records), indexing is also made at more specific levels such as 'Cytochromes' (over 5000 records) and 'Transferrins' (over 17,000 records).

Finally, common names and Enzyme Commission (EC) numbers for enzymes are present in REGISTRY and the substances are easily found through **Substance Identifier**. In other cases, the substrate name is combined with an action term (e.g. Dehydrogenase, lactate) or the catalytic mechanism (e.g. Proteinase, serine).

---

*Protein Nomenclature*

Protein nomenclature and indexing is subject to a number of policies, and the notes above are given not so that SciFinder users will attempt to search complicated names, but rather so that users will be able to interpret names when they appear in REGISTRY records or as Index Terms under **Categorize**.

---

## 6.10   Searching for Information on Polymers

Polymer chemists will be familiar with terms used in their field by authors, and these terms often appear in the titles and abstracts of articles (and hence also in these fields in databases). However, the terms do not always appear and different authors may use a host of different terms for the one topic. Some polymers are known by well over 100 different names, so searches on author-derived terms can be challenging.

Thus it is important to understand how topics and substances in the field are *indexed*. An example of the indexing in a record (AN 2007:1173834) is shown in Figure 6.14.

There are several points to note under '**Concepts**' in the left-hand column, including:

- Substance Class Headings for polymers are shown. There are several hundred headings of this type in CAPLUS, which range from very general headings such as Polyesters

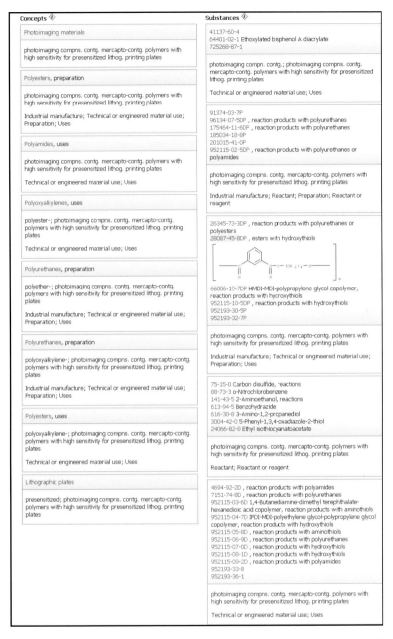

***Figure 6.14*** *Indexing section of the record AN 2007:1173834 in CAPLUS. This shows indexing with polymer Substance Class Headings and with CAS Registry Numbers, and shows an example of indexing of post-treated polymers. SciFinder® screens are reproduced with permission of Chemical Abstracts Service (CAS), a division of the American Chemical Society*

(over 240,000 records) and Polyoxyalkylenes (over 123,000 records) to more specific headings such as Polyoxymethylenes (over 13,000 records). Each substance class has a precise definition, e.g. Polyoxyalkylenes are defined as 'polymers containing only repeating oxyalkylene linkages or segments of repeating oxyalkylene linkages in the backbone';

- These Substance Class Headings are linked to further information, and if a link is clicked then all records in CAPLUS that contain this Index Heading are retrieved (e.g. in the case of Polyoxyalkylenes then over 123,000 records will be retrieved, and these can be narrowed further);

- CAS Roles are associated with Substance Class Headings. If Substance Class Headings with specific CAS Roles are required, then one option is to obtain the appropriate answer set first and then **Analyze: Index Terms**. The Headings and CAS Roles appear in the histogram. Another option is **Explore References: Research Topic** 'polyoxyalkylenes with uses', which will retrieve, among other things, records with the term 'uses' associated with the Heading 'polyoxyalkylenes';

- Many Substance Class Headings have Subheadings; e.g. Polyoxyalkylenes, polyester-; appears in Figure 6.14. If this is a topic of interest, then **Explore References: Research Topic** 'polyoxyalkylenes with polyester' may be considered as a starting point.

There are different points to note under '**Substances**' in the right-hand column, including:

- Some CAS Registry Numbers (e.g. 41137-60–4) do not have associated names so it is important to use CAS Registry Numbers as search terms. (Further comments about this substance appear below.);

- Several CAS Registry Numbers are followed by 'D', which in general indicates that a derivative (sometimes of unspecified structure) of the substance is reported. In this case, the D suffix is used to indicate a post-treated polymer and the entry '952115-07-0D, reaction products with hydroxythiols' is typical of the indexing of post-treated polymers. Hence one option to search post-treated polymers is to enter the CAS Registry Number with 'reaction products' under **Explore References: Research Topic**. The answers may not be comprehensive, but they will show specific examples and the search may be revised with important terms now observed;

- Substance 28087-45-0 (Appendix 4, Section A4.4.3) appears, which is an example of indexing of polymers as a 'structure repeating unit'.

Figure 6.14 shows the types of information relating to polymers in CAPLUS, and the notes above refer to one indexing policy for polymers in REGISTRY. Thus when the actual structure of a polymer is clearly understood through the chemistry of its formation, the polymer is indexed as the product, and 28087-45–8 (the polymer from isophthalic acid (**3**) and 1,4-butanediol (**4**)) must have a single structure repeating unit (**5**). (The chemical process can only produce substances of this part-structure.) In these cases the molecular formula consists of the elements in the structure repeating unit, followed by the subscript 'n' (Section A4.4.3).

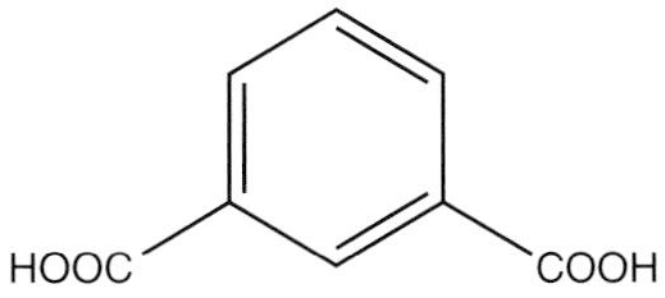

(3)

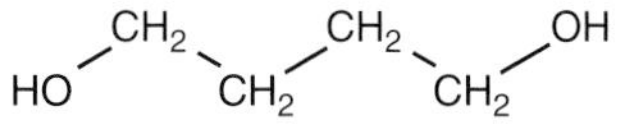

(4)

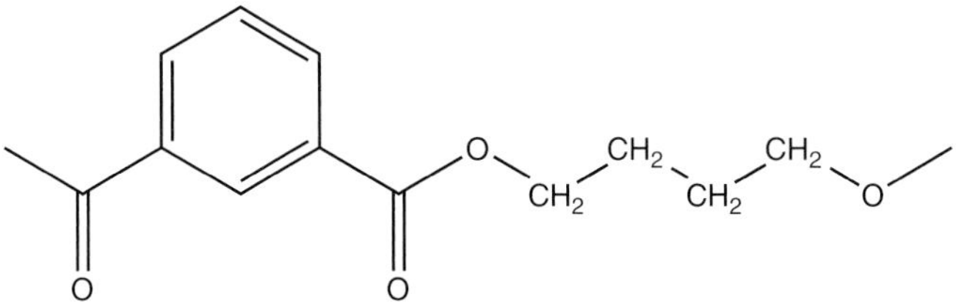

(5)     Molecular Weight     $(C_{12}H_{12}O_4)_n$

Some polymers of this type are registered *only* as structure repeating units, e.g. nylon 6 (CAS Registry Number 25038-54-4), nylon 66 (32131-17-2), polyethylene glycol (25322–68-3), and polyethylene terephthalate (25038-59-9). However, in many other cases, the registration of polymers of this type is a *supplemental registration* and the polymers are registered primarily as *monomer-based substances*. Indeed, the primary registration of copolymers is as the monomer components, and an example is given in Section A4.4.2. In these cases the entry in the molecular formula field relates to the molecular weights of the monomers, and the subscript 'x' follows.

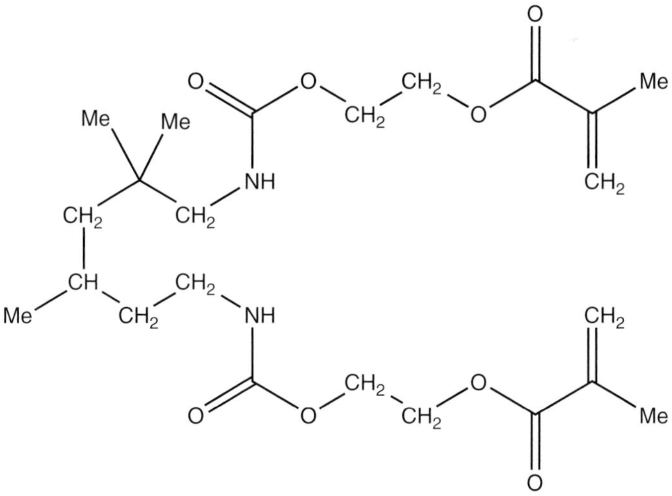

(6)

The substance with CAS Registry Number 41137-60–4 (Figure 6.14) is the bis-methacrylate (**6**). This substance may be found either by name, by formula, or probably most readily by an exact structure search. Once the structure is drawn, **Exact search** is chosen and then in the screens that follow either:

- The **Characteristic: Single component** may be chosen to find the monomer; or
- The **Class(es): Polymers** may be chosen to find copolymers in which the bis-methacrylate (**6**) is one of the components.

As an example of how the search may progress, the latter option gives over 400 polymers and **Get References** retrieves over 200 references. These may now be post-processed with the usual tools; one of the displays after **Categorize** is chosen is shown in Figure 6.15. Applications of these polymers for dental devices are evident. However, this is just an example to show the opportunities that are available in SciFinder for polymer chemists.

---

*SciFinder Tip*

Combinations of search options are often needed for effective searches of polymers. There are several types of indexing, and while it is helpful to know the basic indexing from the outset, nevertheless it is not essential. Rather what is essential is that the user looks through records, identifies important indexing, and then implements this knowledge to construct new strategies.

For a summary of polymer search strategies on SciFinder see links in Appendix 1.

---

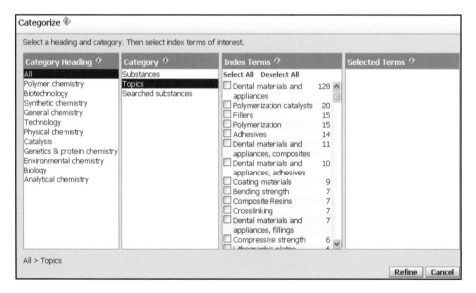

**Figure 6.15** *One of the many **Categorize** displays for bibliographic records that report information on polymers that contain the bis-acrylate (6). SciFinder® screens are reproduced with permission of Chemical Abstracts Service (CAS), a division of the American Chemical Society*

## 6.11   Summary of Key Points

- A single author may be represented in several different ways in CAPLUS and in MEDLINE and a single name may refer to several authors. The various entries may be identified through **Explore: Author Name** and **Analyze: Author Name**;
- Even more variation occurs with entries for names of companies and both **Explore: Company Name** and **Analyze: Company Name** should be considered. It is better to enter only parts of the company name initially and then to make further decisions based on the information obtained;
- **Explore: Document Identifier, Explore: Journal**, and **Explore: Patent** offer options to identify records or types of documents;
- From the references screen, **Get Substances, Cited References**, and **Citing References** offer ways to broaden searches;
- Initial routes to find information on substances may use either **Explore References: Research Topic** or **Explore Substances**. Each has particular applications, although often it is advisable to look at both options;
- The bibliographic databases MEDLINE and CAPLUS, and their links to REGISTRY, provide enhanced opportunities for those who traditionally use only MEDLINE as their source of biomedical information. SciFinder search intelligence and post-processing functions, particularly **Analyze** and **Categorize**, assist further to achieve comprehensive and precise answers;
- Searching for substances in the biological sciences and for polymers is particularly challenging, partly because the names of substances may appear in original documents in many different ways. It helps to have an understanding of the indexing of these substances beforehand. More effective searches may be implemented at the outset, but then it is still necessary to look through records to ensure that the most appropriate search terms have been used.

# 7

# Searching for Information on Chemical Reactions

## 7.1   Introduction

There are numerous ways in which reactions are described and numerous reasons why they are performed. Reactions may be described by their type (e.g. oxidation, reduction, addition, elimination, substitution, polymerization, cyclization, metathesis) or by their intent (e.g. to study synthetic methods, to prepare substances, or to undertake a mechanistic study). Chemists also describe reactions by their names (e.g. Wittig reaction, Dess–Martin oxidation, hydroboration, Grob fragmentation) or by acronyms (e.g. IMDA – intramolecular Diels–Alder). Substances involved may be described as reactants, reagents, products, or catalysts. Other factors relating to reactions include conditions such as solvents, temperature, photochemical reactions, and yields.

The various terms that describe reactions are used extensively by *authors* in titles and abstracts in bibliographic databases, and a *wealth of chemical reaction information appears in these fields in CAPLUS*. Chemical reaction information is also entered by CAS *document analysts*, primarily as Index Headings and through the CAS Roles (Section 2.1.2 and Appendix 2) associated with CAS Registry Numbers or Substance Class Headings. Additionally, CAS has created a separate chemical reaction database (CASREACT, Section 2.4) in which reactions are fully indexed with atom-by-atom mapping, bonds formed/broken in the reaction, and other parameters including type of reaction and yield. There are a very large number of ways to describe, and hence also to find, information on reactions!

*Information Retrieval: SciFinder®, Second Edition* Damon D. Ridley
© 2009 John Wiley & Sons, Ltd

*Search Note*

Chemical reaction information may be challenging to search directly in full text documents, particularly where information is often presented in reaction diagrams that are not searchable. Information may also appear in text format, but author terms vary considerably.

Chemical reaction databases are designed to overcome these problems and they allow precise reactions to be explored with structure-based queries, but it helps to remember that there are many other ways to search reactions. For example, CAPLUS offers options through searching terms used by authors in titles and abstracts, through CAS Registry Numbers and CAS Roles (particularly PREP and RCT), and through reaction Index Headings (e.g. Suzuki Reactions and Suzuki Reaction Catalysts).

Different options need to be used to retrieve different types of reaction information. The information comes from different databases but it is simple to transfer information from one database to another on SciFinder. SciFinder post-processing tools offer systematic ways to narrow or broaden searches.

While authors may enter reaction information in titles and abstracts, often reaction information may appear only in the text of the full article. Consider, for example, the bibliography and abstract in the article by Zhang and Breslow (Appendix 6) and the series of reactions in Schemes 1 and 2 in this original document. These reactions are not mentioned in the title or the abstract, but Schemes 1 and 2 are indexed both in CAPLUS and CASREACT. For example, Figure 7.1 shows the sequence in Scheme 1 (Appendix 6) together with the CAS Registry Numbers of the substances involved and the CAS Roles associated with these Registry Numbers *in the bibliographic database*. In the record, 42 CAS Registry Numbers are indexed with the role PREP and 47 CAS Registry Numbers with the role RCT. For details of the indexing, the record (AN 126:183046) in SciFinder may be viewed. Scheme 2 is also fully indexed in a similar way in this database.

On the other hand, the record for the original article in the *reaction database* is quite different. This database contains information only on Scheme 2 (there are a number of indexing policies applied in the construction of reaction databases, and generally for older records in CASREACT only key reactions are entered) and the six synthetic steps from 2-bromopyridine (**11**) to the substituted cyclodextrin (**10**) are indexed as six different reactions (Figure 7.2). However, in order to allow for questions that involve multistep reactions, 21 'reactions' appear in the record in the database for this sequence; i.e. the two step sequence (**11**) to (**13**), and each of the other multistep reactions from (**11**) up to the six-step sequence (**11**) to (**10**) are indexed as 'reactions', together with the various multistep processes from synthetic intermediates (**12**) through (**16**).

It is apparent from the above example that there are many fundamental issues to consider when setting up the reaction query. First, reaction information may be searched in several ways, which are summarized in Table 7.1 together with a summary of the search process and database coverage.

Second, the CAPLUS bibliographic database covers more than 30 million documents from 1907 plus more than 134 000 pre-1907 records. While searches in the title and abstract fields may lack precision, more precise information on reactions and preparations in CAPLUS may be retrieved through searching CAS Registry Numbers and linking them

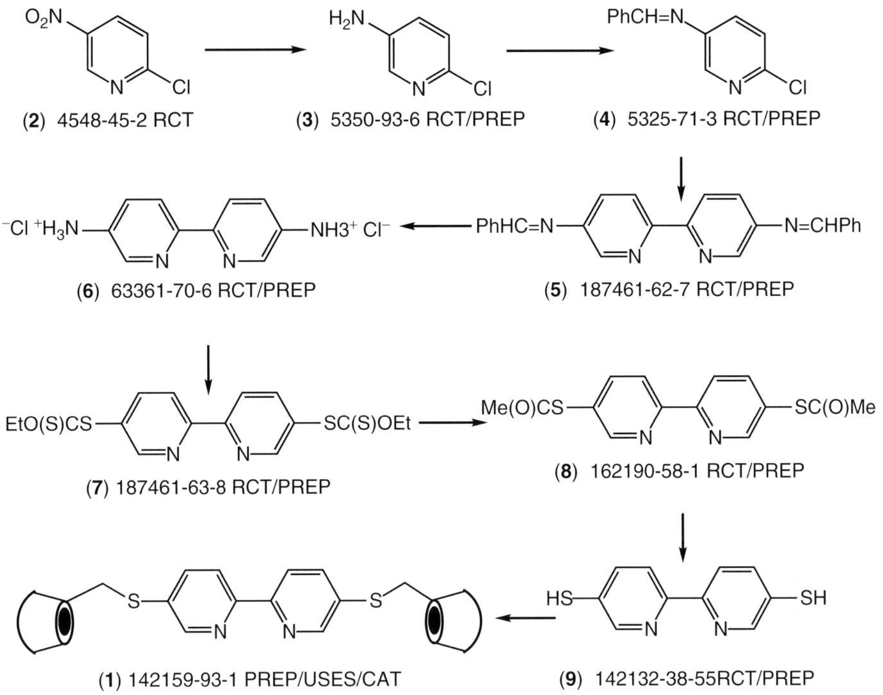

***Figure 7.1*** *Indexing in CAPLUS of the reaction sequence in Scheme 1 (Appendix 6). Substance numbers are those in the original publication. The CAS Registry Numbers and CAS Roles entered in the CAPLUS record are shown*

with CAS Roles. Currently the PREP and RCT roles occur in more than 4.7 million and 3.2 million of the records respectively, and, since these may appear many times in a single record, the numbers of preparations and reactions listed in this way is very extensive.

Third, searches in the reaction database may be refined with great precision, but fewer original articles have records in the reaction database, and it helps if indexing issues are understood.

## 7.2   Specific Search Options in CASREACT

Searches in CASREACT are commenced by clicking **Explore Reactions** in the SciFinder main screen. After the initial screen (Figure 7.3) appears, the structure editor box is clicked and the query is drawn in the reaction editor (Figure 7.4).

The screen is similar to the structure editor (Figure 4.2), except for five additional drawing tools (Table 7.2) and the two search options on the right of the screen.

**Map atoms** ❸ and **Mark bonds** ❹ are tools that are used to define the conversion more precisely. For example, if the user requires references to reactions of the type shown in Scheme 1 (Figure 7.5, i.e. perhaps Wittig type reactions involving $Ph_3P = CHOMe$), one option would be to draw the scheme exactly as shown. SciFinder looks for all reactions in CASREACT that have the cyclohexanone substructure as a reactant

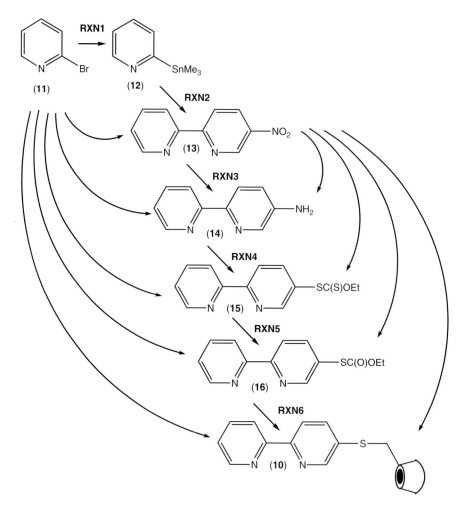

**Figure 7.2**  *The six reactions in Scheme 2 (Appendix 6) are indexed as 21 'reactions' in CASREACT in order to allow for searches on multistep reactions*

and the part-structure shown in the product. Many answers will be of the required type (e.g. reaction A), but reactions B, C, and D will also be retrieved since they meet the search requirements. However, these last three reactions are of quite a different type from the required transformation.

There are many ways to proceed. One would be to request that the double bond in the starting material be 'formed or broken' (this is done through the **Mark bonds** icon), and now reactions of types A and B would be retrieved. Another would be to request that the double bond in the product be 'formed or broken' (reactions of types A and C would be retrieved), while if the double bonds in both reactant and product were tagged then reactions of type A only would be retrieved.

Additional precision may be obtained by mapping atoms in the reaction. However, the search process in SciFinder is quite precise and works differently from other commonly

**Table 7.1**   *Options to search for reaction information in SciFinder*

| Search Option through SciFinder | Search Process | Coverage |
|---|---|---|
| **Explore References: Research Topic** or **Refine References: Research Topic** | Words are searched in title, abstract, and indexing in CAPLUS. | CAPLUS >10,000 serials and patent documents from 57 patent authorities. Coverage from 1907–. Includes information from *Organic Reactions, Organic Synthesis*, and *EROS* (*Encyclopedia of Reagents for Organic Synthesis*). |
| **Explore Substances** (e.g. by **Chemical Structure**) | Substances are found in REGISTRY. | REGISTRY Coverage since 1957 but many substances back to early 1900s. |
| Then **Get References** (References associated with **Preparation** or **Reactant/Reagent**) | Answers are found in CAPLUS and have CAS Registry Numbers closely associated with the CAS Roles PREP or RCT. | CAPLUS There are 4.7 million and 3.2 million records in CAPLUS that contain the PREP and RCT roles respectively. However, a single record may have these roles assigned to many substances and the actual number of substances involved may approach 100 million. |
| **Explore Reactions** | Reactions are found in CASREACT. | CASREACT Selected serials in synthetic organic chemistry from 1840–; some patents from 1991–; information from *Organic Reactions, Organic Synthesis*, and *EROS*. |

used reaction databases, which often require extensive mapping of atoms/bonds to obtain reasonably precise answers. In SciFinder it is generally sufficient to map only one or two atoms or bonds and the mapping shown in Figure 7.4 is sufficient to obtain precise answers in this case.

## 7.3   Reaction Search Strategies

There are a number of questions to consider before the search is conducted. For example, one of the key reactions in the article by Zhang and Breslow is the Stille reaction (Figure 7.6), and at issue is how to proceed to find this and related reactions. Note that no mention of this reaction is made in the title or in the abstract, so on this occasion searches on words in these fields will not retrieve this record and the searcher must rely on terms entered by the document analyst (in CAPLUS, CASREACT, or REGISTRY).

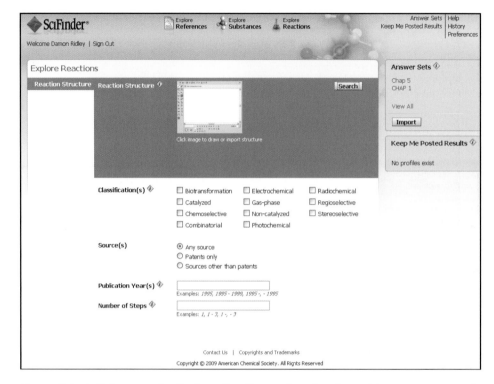

**Figure 7.3**  *Initial screen for **Explore Reactions**.  Searches from this screen are run in CASREACT and usually start with structure queries. Normally it is better to choose additional search options later through **Analyze/Refine** of initial answer sets. SciFinder® screens are reproduced with permission of Chemical Abstracts Service (CAS), a division of the American Chemical Society*

The simplest search strategy would be to draw, then search, a query as shown in Figure 7.6. When this is done the Zhang and Breslow reaction is the only answer found. However, synthetic chemists usually need to consider analogous reactions and a number of questions arise, including:

- What variations would be acceptable?  For example, would a preparation from the pyridyl 2-tri*butyl*stannane and 2-*bromo*-5-nitropyridine be acceptable?
- If the exact reaction or simple variations are not known, then what type of information may be acceptable?  Would information on the preparation of a possible precursor (e.g. the aminobipyridine rather than the nitrobipyridine) be acceptable, since a subsequent conversion may be possible to give the new substance?
- Does a crucial aspect of the question really relate to finding reactions in which a bond is formed between two pyridine rings in the 2-position?
- If attempts to answer the above questions in the reaction database do not produce the type of information required, then how may the problem be solved in the bibliographic database?

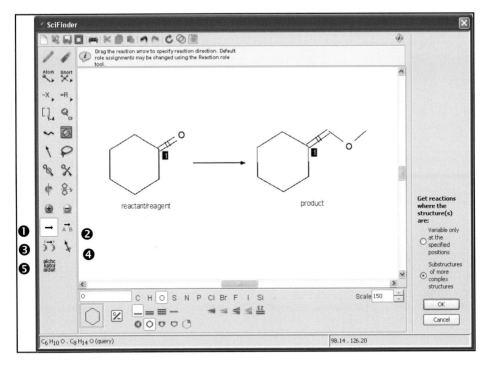

**Figure 7.4**   *Reaction editor screen in which the reaction query is built. The numbers inserted in this figure are referred to in Table 7.2. SciFinder® screens are reproduced with permission of Chemical Abstracts Service (CAS), a division of the American Chemical Society*

**Table 7.2**   *Summary of functions referenced in Figure 7.4. Further information on these functions is described in this chapter*

| Locator in Figure 7.4 | Function | Brief description |
|---|---|---|
| ❶ | **Reaction arrow** | Drag the reaction arrow to specify the direction of the reaction. Reactant/reagent and product roles are then inserted. |
| ❷ | **Reaction role** | Reaction roles may be added to each structure or functional group. Options are: product, reactant, reagent, reactant/reagent, any role. Additional option for functional group: nonreacting. |
| ❸ | **Map atoms** | Map atoms in starting materials and products. |
| ❹ | **Mark bonds to be formed or broken** | Mark bonds that are changed in the reaction. |
| ❺ | **Functional group** | Choose a functional group from a list (which is displayed when the tool is clicked). |

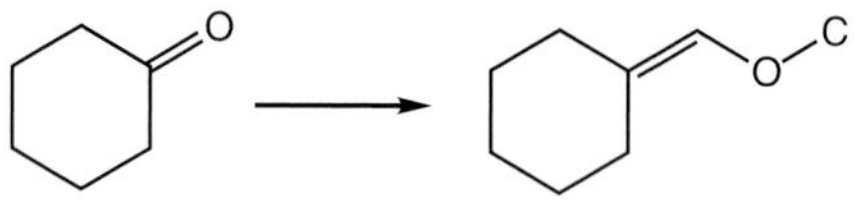

Scheme 1

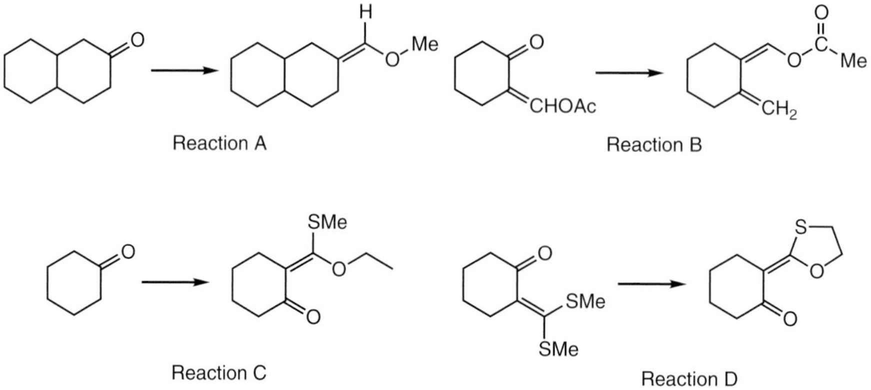

Reaction A

Reaction B

Reaction C

Reaction D

**Figure 7.5**  *Examples of answers found in the search on the reaction query shown in Scheme 1. Marking bonds to be formed or broken may be used to narrow answers to more specific transformations*

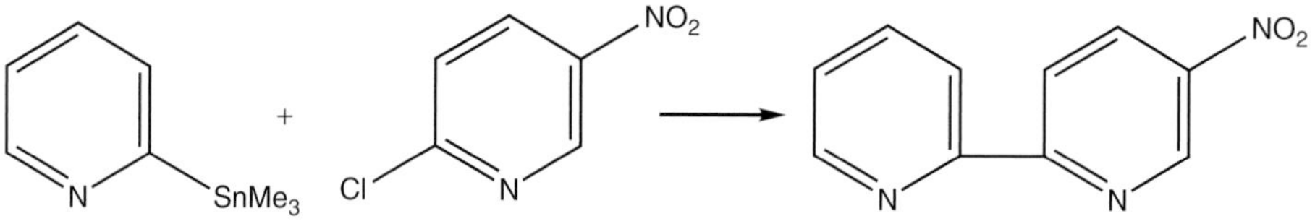

13737-05-8                    4548-45-2                    14805-00-6

**Figure 7.6**  *Stille reaction from the Zhang and Breslow publication (Appendix 6)*

It really does not matter whether the searcher *first* approaches the problem from the most specific option (searching precise reactions in CASREACT) or from the most general option (searching CAPLUS for preparations of products, or reactions involving starting materials, or searching for substances in REGISTRY). It is more important that the searcher realizes there are a number of different approaches and that a few of them are attempted. More general searches are not only more likely to retrieve specific processes of interest but are also likely to retrieve alternatives that the searcher may not have considered initially.

### 7.3.1  Explore Substances and Explore Reactions

The question is 'How can the search be broadened and what types of answers are obtained using various options? Some alternatives are discussed in the section that follows, where the intent is to show experienced synthetic chemists what may be accomplished.

### 7.3.1.1   Explore Substances (i.e. Start in REGISTRY)

At a reasonably general level, the structure of 3-nitrobipyridine (14805-00-6) may be built, and a full substructure search gives all substances with the nitrobipyridine part structure. When **Get References** is chosen (followed by **References associated with: Preparation**), records in CAPLUS that have any of the CAS Registry Numbers from the substructure search closely associated with the CAS Role PREP are retrieved. Such a search currently gives 38 substances (when **Precision analysis: Conventional Substructure** is chosen), which lead to 28 references that describe preparations.

If the chemist is interested only in the synthesis of 3-*nitro*bipyridines then this answer set may meet requirements. Certainly as there are very few answers in this broader search, the chemist probably would not need to proceed to CASREACT. However, the chemist may also be interested in related substances, and particularly in syntheses involving the formation of bipyridines from two separate pyridine units. More general searches now need to be undertaken.

There are many ways to proceed, and a summary of some of the options is given in Table 7.3. Not all of these would be tried, but they are described here simply to show the outcomes of the various approaches and to alert chemists to the additional issues that may arise. A more complete description of the thought processes behind the entries in the table is given below.

The intention of Entry 1 (Table 7.3) is to find all substances with the bipyridine substructure in REGISTRY (i.e. start with **Explore Substances**) and then to find preparations for these substances in CAPLUS. However, an immediate problem is encountered (the search does not complete; see Section 5.2.2) and so the rings are locked (Entry 2). While this excludes all fused rings, still over 17,000 substances are retrieved. The initial answer set automatically displays an analysis of the roles of the substances and it is found that over 11,000 and over 6000 substances have preparation and reaction roles respectively in CAPLUS. The search now probably is too general, and the conclusion is that initial broad searches in REGISTRY in this instance may not be worth exploring further.

### 7.3.1.2   Explore Reactions

The remaining entries all start with **Explore Reactions**; i.e. the searches start in CASREACT. In the first of these (Entry 3), the bipyridine structure is built, the rings are not locked, and the role Product is assigned. Over 74,000 reactions are retrieved (Figure 7.7).

While this is far too many reactions to investigate, already the default screen shows **Analysis: Catalyst**, and a chemist working in this area would immediately recognize that the top few entries (involving palladium and copper catalysts) are of interest.

### 7.3.1.3   Analysis/Refine (Reactions)

The **Analysis** options in screens for CASREACT answers are shown in Figure 7.8. Several of these are based on similar bibliographic fields to those presented in screens for CAPLUS answers (Table 3.4). For example, if certain journal titles are more readily available then answers may be limited to these titles by working through screens for **Analysis: Journal Name**.

**Table 7.3**    *Search options in the substance and reaction databases in SciFinder for different reactions relating to the synthesis of bipyridines*

| Entry | Search query built | Search request | What SciFinder does | Number of answers |
|---|---|---|---|---|
| 1 | **(Explore substances)** (rings are not locked) | 1. Substructure search (check **Precision analysis**) | The query is too general and SciFinder does not complete the search. | |
| 2 | **(Explore substances)** (rings are locked) | 1. Substructure search (check **Show precision analysis**) 2. **Analyze: Substance Role** | A substructure search of the query in REGISTRY. **Analyze: Substance Role** is the default. | 17,307 substances (Conventional substructure) Analysis states 11,586 and 6831 of the substances have PREP and RCT Roles respectively in CAPLUS. |
| 3 | **(Explore reactions)** product (rings are not locked) | 1. **Substructures of more complex structures** 2. **Refine reactions** with structure of Sn (Reactant/ reagent) | Performs a substructure search in CASREACT and retrieves substances mapped with role PREP. Narrows answers to those that have Sn in the structure of the starting material. | 74,296 reactions (see Figure 7.7) 4870 reactions (but many describe reactions of tin-substituted biphenyls) which are reported in 324 references. |
| 4 | **(Explore reactions)** product (rings are locked) | 1. **Substructures of more complex structures** 2. **Refine reactions** with structure of Sn (Reactant/ reagent) | Performs a substructure search in CASREACT and retrieves substances mapped with role PREP. Narrows answers to those which have Sn in the structure of the starting material. | 19,061 reactions. 3130 reactions which are reported in 166 references. |

**Table 7.3**  *(continued)*

| Entry | Search query built | Search request | What SciFinder does | Number of answers |
|---|---|---|---|---|
| 5 | (Explore reactions)<br><br>product<br><br>(rings are locked) | 1. **Substructures of more complex structures**<br><br>2. **Refine reactions** with structure of Sn (Reactant/ reagent)<br><br>3. **Analysis: Product Yield** (> = 90 %) | Similar search to Entry 4 except only answers with the tagged bond formed in the preparation are retrieved.<br><br>Narrows answers to those which have Sn in the structure of the starting material.<br><br>Narrows answers to those reported to proceed in > = 90% yield. | 4591 reactions, which are reported in 395 references.<br>2404 reactions, which are reported in 103 references.<br>(See Figure 7.11.) |
| 6 | (Explore reactions)<br><br>reactant<br><br>product<br><br>(rings are locked) | 1. **Substructures of more complex structures** | Similar search to Entry 5 except that only answers with the 2-chloropyridine substructure as reactant are retrieved. | 846 reactions, which are reported in 78 references.<br><br>If X rather than Cl is used in query, then 3607 reactions which are reported in 280 references are retrieved. |
| 7 | Reactant + Reactant | 1. **Substructures of more complex structures** | Searches all records in CASREACT with CAS Registry Numbers from substructure searches mapped with role RCT. | 1876 reactions, which are reported in 101 references. |

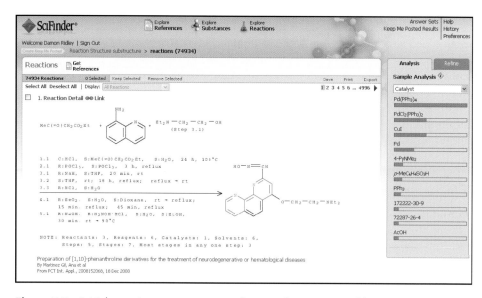

**Figure 7.7**   *Initial reaction answer screen for search Entry 3, Table 7.3.   Over 74,000 reactions are retrieved, but already information of interest is displayed through **Sample Analysis: Catalyst** on the right of the screen.   This or other **Analysis** or **Refine** options now need to be used to narrow answers.  SciFinder® screens are reproduced with permission of Chemical Abstracts Service (CAS), a division of the American Chemical Society*

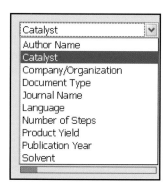

**Figure 7.8**   **Analyze** *options available for CASREACT answers.   Six options are based on bibliographic entries; the remaining four options relate specifically to reaction information. SciFinder® screens are reproduced with permission of Chemical Abstracts Service (CAS), a division of the American Chemical Society*

The options **Catalyst, Number of Steps, Product Yield**, and **Solvent** are specific to CASREACT. Thus one way to narrow answers is to retrieve only those that are reported to proceed in $>= 90\%$ yield or perhaps only those performed in certain solvents (e.g. for environmental reasons, in water). These options are valuable, but interpreting the data for multistep reactions needs to be considered carefully.

*SciFinder Note*

Single-step reactions are generally defined as 'one-pot reactions'. However, the distinction between single-step and multistep reactions may be blurred. For example, consider the one-pot reaction where an alkyl halide is added to magnesium and then a carbonyl compound is added after the formation of the Grignard reagent is complete.

Because of this complication and because SciFinder has slightly different ways of linking reagents in single-step and in multistep reactions, careful interpretation of results is needed at times.

The **Refine** options, together with the specific inputs available, in screens for CAS-REACT answers are shown in Figure 7.9. Their application is intuitive, although note:

- **Product Yield** shows a box **Include answers that have no product yield**. Yields for reactions are not always reported in the original document and this box enables such reactions to be included in answers. However, if high yielding reactions are the principal requirement then the user may not want to include this option;
- As indicated in the SciFinder Note above, care needs to be exercised in interpreting the **Number of Steps** for multistep reactions. The best solution is to work through some answers and to understand issues before taking further action;
- **Reaction Classification** refers to broad reaction categories; its use depends on the intent of the query. The best solution is to work through some options.

Accordingly, there are many ways to proceed from the initial answer set of over 74,000 reactions (Figure 7.7). If reactions of the type shown in Figure 7.6 (Stille reactions) are of most interest then one general option may be to **Refine: Chemical Structure** and to build a query that simply contains a single Sn atom (which is specified as a reactant/reagent). Entry 4 (Table 7.3) shows the outcome of first locking the bipyridine rings and then limiting reactions to those that contain an Sn atom. Figure 7.10 shows the first of the 3130 reactions retrieved.

This answer matches the search intent, but a glance through some of the *other* answers indicates that many involve organotin compounds of bipyridines, i.e. subsequent

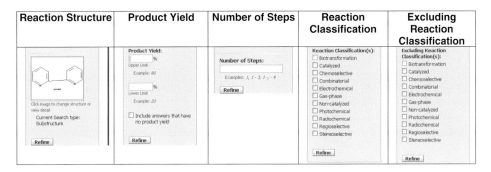

**Figure 7.9** **Refine** *options available for CASREACT answers. SciFinder® screens are reproduced with permission of Chemical Abstracts Service (CAS), a division of the American Chemical Society*

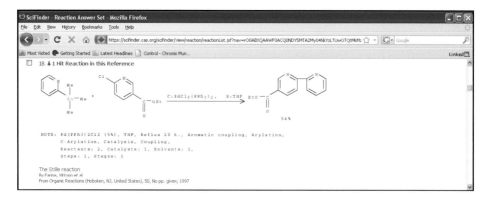

**Figure 7.10**   *An answer obtained when the answer set shown in Figure 7.7 is refined with the structure query: Sn (reactant/reagent). See Entry 4 in Table 7.3 for a further summary. First listed answer is from Organic Reactions. SciFinder® screens are reproduced with permission of Chemical Abstracts Service (CAS), a division of the American Chemical Society*

chemistry of bipyridines and not chemistry related to the required formation of the bipyridine bond. While it is not easy to forecast such problems or to know the extent of them in advance, nevertheless it is a simple matter to try the general experiment. Those involved with organic synthesis know that often a quick way to optimize the chemistry is to conduct reactions under 'extreme' conditions first; if these do not work, then more gentle reaction conditions are tried. Thus the thought processes in the various options in Table 7.3 follow the experimental approaches that a synthetic chemist would apply in the laboratory!

### 7.3.1.4   Marking Bonds

To remove answers that involve subsequent chemistry of bipyridines, the key bond of interest is tagged (Entry 5 in Table 7.3). When the initial 4591 answers are narrowed to include those with tin in the reactant, then 2404 reactions are retrieved. However, they are reported in only 103 references. **Analysis: Product Yield**, then restricting answers to those with $>= 90\ \%$ yield, affords 10 answers (Figure 7.11). Note that this answer would *not* have been retrieved had a query of the type shown in Figure 7.6 (tin atom attached to a pyridine ring) been searched. Nevertheless, it may be of interest where symmetrical bipyridines were required and it may suggest other approaches to solve the problem at hand.

Entry 6 (Table 7.3) moves to a greater level of precision (a 2-chloropyridine is specified as a starting material) while Entry 7 gives another option (reaction between a 2-halopyridine and a 2-stannylpyridine). This last Entry gives 1876 reactions in 101 references. On this occasion the first answer that appears is a publication from 1997 (shown in Figure 7.10).

Records from *Organic Reactions, Organic Synthesis*, and *EROS* have been added recently to CASREACT and CAPLUS, and the records contain full CAS indexing. For example, the record in CAPLUS for the review article (Figure 7.10) contains almost 6000 CAS Registry Numbers, i.e. indexing for all reactions in the review. If this record

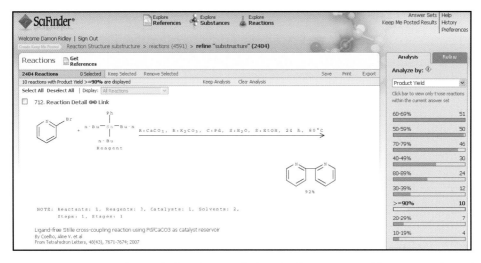

**Figure 7.11**   *Screen obtained when the search described in Entry 5 (Table 7.3) is performed and when answers are restricted to those that are reported in >= 90 % yield. SciFinder*® *screens are reproduced with permission of Chemical Abstracts Service (CAS), a division of the American Chemical Society*

(AN 2008:1383562) is obtained in CAPLUS and then **Get Substances** is chosen, then all the indexed substances are displayed, which may be analysed or refined with any of the tools described earlier for substances (Section 5.3).

In summary, Table 7.3 shows a number of options for searching reaction information ranging from general to quite specific queries. A general search immediately showed some interesting catalysts, and a more specific search showed a different approach (although probably limited to the synthesis of symmetrical biphenyls).

### 7.3.2   Using Functional Groups

While chemists are often interested in the reactions of substances of a certain structures, sometimes the key requirement is simply the chemistry of a particular functional group. Of course functional groups may always be represented by part-structures, and one approach to find methods to convert 1,2-diols (glycols) to aldehydes may be to search for the reaction in Figure 7.12.

To do this, SciFinder would have to find all reactions that have the glycol substructure in the starting material and the aldehyde substructure in the product. However, such a

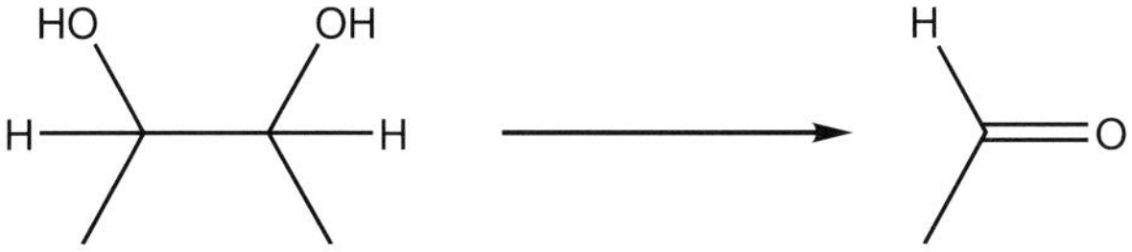

**Figure 7.12**   *Structure query for search for methods to convert glycols to aldehydes*

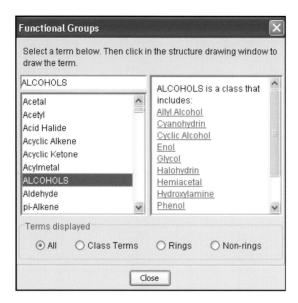

**Figure 7.13**  *Screen obtained when Functional Group icon (locator ⑤ in Figure 7.4) is clicked.  Click on functional group class (left column) and specific examples of the class appear (right column).  SciFinder® screens are reproduced with permission of Chemical Abstracts Service (CAS), a division of the American Chemical Society*

general search would exceed the system search limits in CASREACT. The solution is to search initially by functional groups.

A large number of functional groups are built into SciFinder and these are displayed by clicking on the functional group icon in the structure drawing screen (Figure 7.13). Clicking on the functional group (left column) displays more specific functional groups in the class (right column). Searches may be performed at the specific or class level. Once the required group has been chosen, clicking on the structure drawing screen enters the functional group term.

In this way a query (Figure 7.14) may be set up and the reaction arrow (→) or the reaction role icon (**A**→**B**) may be used to increase precision. Actually when the latter icon is clicked on a functional group in the structure drawing screen, there are five choices (reagent, reactant/reagent, product, any role, and nonreacting). When **Get Reactions: Substructures of more complex structures** is selected the result is an answer set of over 19 000 reactions. This answer set may then be analysed or refined with any of the tools mentioned in this chapter.

A second important application of searches by functional group is the use of the 'nonreacting' option, which is particularly important when queries about *selective reactions of functional groups* are required. For example, to search for reagents that will selectively oxidize sulfides to sulfones in the presence of other oxidizable groups (e.g. alcohols), it is easy to set up the functional group query: sulfides (reactant/reagent), sulfones (product), alcohols (nonreacting).

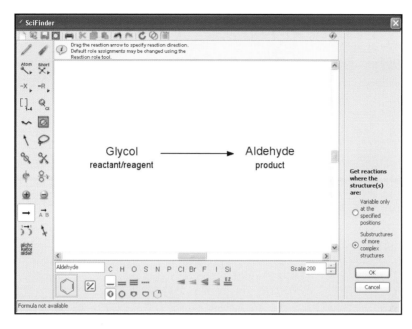

**Figure 7.14**  *Functional group query for the conversion of glycols to aldehydes. SciFinder®*
*screens are reproduced with permission of Chemical Abstracts Service (CAS), a division of*
*the American Chemical Society*

Modern synthetic chemistry commonly involves reactions of substances with a variety
of functional groups, but where reaction of only one of them is required. Use of the
'nonreacting' option solves this in a simple way. For example, if the actual glycol
cleavage to be performed is on a substance that contains a phosphonate group, then the
query (Figure 7.15) may be searched; one of the 124 hit reactions is shown in Figure 7.16.

If 'any role' is assigned to the phosphonate in the query (Figure 7.15) then over
1800 hit reactions are obtained, but most of these are multistep reactions where the
phosphonate is involved in quite a different reaction from the glycol cleavage reaction.
This example illustrates the need to interpret results from multistep reactions carefully,
and in many cases it may be necessary to narrow hits to single-step reactions.

In summary, searching by functional groups helps overcome system limits for general
reaction searches and also allows for the very useful option of finding reactions that
are selective for one type of functional group over another. Such searches may be
conducted at any stage, i.e. either as an initial search or under **Refine Reactions**, and
in turn answers may be analysed or refined (e.g. with a more precise structure such as
cyclohexane-1,2-diol if the main interest is in glycols of this type).

### 7.3.3  Retrosynthetic Analysis

To set up retrosynthetic analyses in CASREACT it is simply necessary to draw the
structure of interest, to mark a bond to be formed/broken in the reaction, and to specify

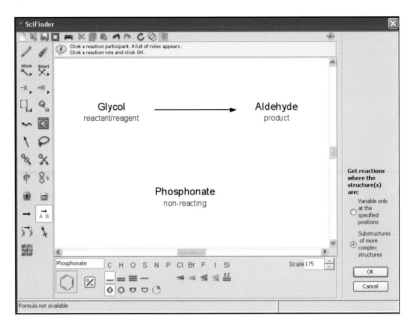

**Figure 7.15**   *The functional group query in Figure 7.14 has been modified so that only reactions also containing a nonreacting phosphonate group are retrieved. SciFinder® screens are reproduced with permission of Chemical Abstracts Service (CAS), a division of the American Chemical Society*

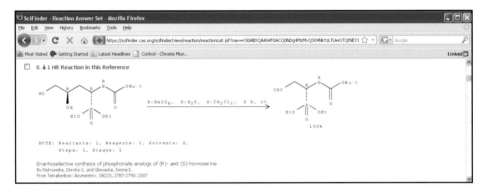

**Figure 7.16**   *Sample answer from the query in Figure 7.15. SciFinder® screens are reproduced with permission of Chemical Abstracts Service (CAS), a division of the American Chemical Society*

the role 'product' for the structure. Different bonds may be marked in turn and hence different retrosynthetic routes may be evaluated.

For example, to perform a retrosynthetic analysis on epibatidine (**17**), the structure is drawn directly in the reaction editor screen. Alternatively, the record for epibatidine in REGISTRY is found (e.g. through **Explore Substances: Substance Identifier** and

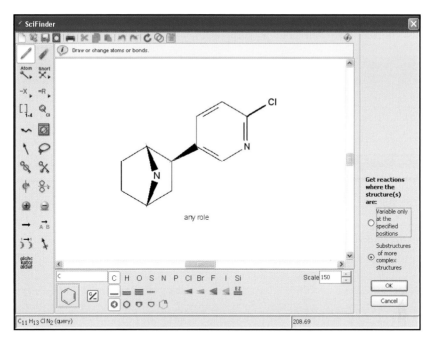

**Figure 7.17**  *Result when structure from REGISTRY is pasted into the reaction editor. Hydrogens may be omitted from the original structure and 'any role' is specified. SciFinder® screens are reproduced with permission of Chemical Abstracts Service (CAS), a division of the American Chemical Society*

entering the name); when the structure is clicked an option to **Explore Reactions** is presented (see Figure 5.10). When this link is chosen, SciFinder automatically pastes the structure (usually without hydrogens attached) into the reaction structure editor and specifies 'any role' (Figure 7.17). Of course this option is particularly convenient when reactions on relatively complex structures need to be searched.

**(17)**

If the reaction role is changed to 'product' and if **Substructure of more complex structures** is chosen, then all preparations in CASREACT will be retrieved. However, if specific bonds are marked to be formed or broken, then specific retrosynthetic pathways may be investigated. For example, if the bonds shown in structure (**18**) and structure (**19**) are tagged, reaction searches give around 250 reactions in each case and Figure 7.18 and Figure 7.19 respectively show examples of answers. Graduate students can thus use SciFinder to surprise their professors with their retrosynthetic skills!

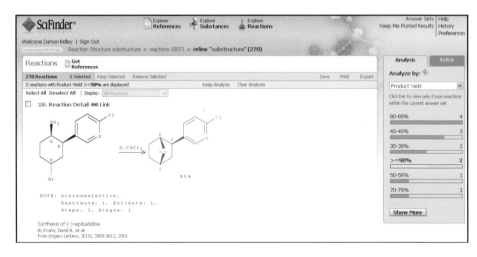

product

**(18)**

product

**(19)**

---

*Computer-Aided Retrosynthetic Tools*

The classic retrosynthetic programs use algorithms to devise retrosynthetic routes. For example, Michael-type reactions offer routes to β-substituted ketones, so the algorithm considers additions to unsaturated ketones in such cases. Further, certain types of reactions are weighted. For example, if a cyclohexene is identified in a possible sequence then Diels–Alder additions will be considered.

On the other hand, retrosynthetic analysis in SciFinder looks at actual reactions performed and answers may then be analysed, for example, by reaction yield. Advanced synthetic strategies can thus be devised when these two approaches are used together.

---

**Figure 7.18**  *Sample answer for retrosynthetic analysis in which the bond in structure (18) is tagged to be 'formed or broken'. Initial answers have been narrowed to those that are reported to proceed in > = 90 % yield. SciFinder*® *screens are reproduced with permission of Chemical Abstracts Service (CAS), a division of the American Chemical Society*

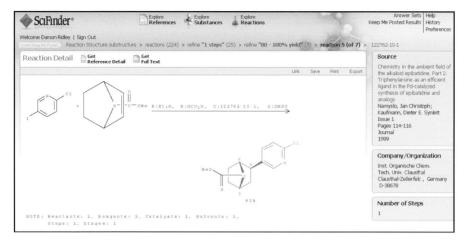

**Figure 7.19** *Sample answer for retrosynthetic analysis in which the bond in structure (19) is tagged to be 'formed or broken'. When the reagent C:122762-10-1 is clicked, the substance (a bis-triphenylarsine palladium complex) is displayed. SciFinder® screens are reproduced with permission of Chemical Abstracts Service (CAS), a division of the American Chemical Society*

## 7.4 Searching for Reactions through Explore References: Research Topic

The many points of entry to information on chemical reactions in CAPLUS are summarized in Table 7.4, and there are occasions when to search for reaction information initially through **Explore References: Research Topic** is preferred.

First, consider options available if information on the reaction of ethyl propiolate with triethylamine is required (Figure 7.20). The most precise option is to draw the starting

**Table 7.4** *Search options for reactions in CAPLUS*

| Section of record | Origin of entry | Search option in SciFinder |
|---|---|---|
| Words in titles and abstracts; Text-modifying phrases | Author[a] | Through **Explore References: Research Topic** or **Refine: Research Topic** |
| Index Headings (for Chemical Reactions) | Document analyst | Through **Analyze: Index Term** or through **Categorize** |
| Supplementary Terms | Document analyst | Through **Analyze: Supplementary Term** |
| CAS Registry Numbers | Document analyst | Preferably through **Explore Substances** (e.g. through a substructure search) |
| CAS Roles | Document analyst | Through **Explore Substances** and then the list of roles that appears at **Get References** (Figure 3.12) |

[a]Titles and abstracts may be modified by the document analyst for patents, and text-modifying phrases based on author terminology are entered by the indexer.

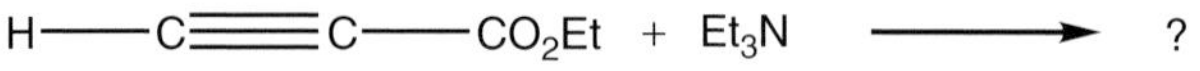

$$H\text{---}C\equiv C\text{---}CO_2Et \;+\; Et_3N \longrightarrow \;?$$

**Figure 7.20**   *What is the product formed after triethylamine is mixed with ethyl propiolate?*

materials in the reaction editor screen under **Explore Reactions**, to specify the structures as reactant/reagent, and then to click **Substructures of more complex structures**.

Another approach is to search functional groups, e.g. to start a search on a query with ALKYNE and TERTIARY AMINE as reactants/reagents. The searcher would need to consider carefully whether to specify the ester group since on chemical grounds it may or may not be involved in the reaction. Yet another approach may be to start with various structure searches in REGISTRY and then to find references where the substances are reported in reactions. In the event, none of these options gives answers that really are satisfactory – mainly because system limits are encountered for such general searches.

On the other hand, when **Explore References: Research Topic** 'reaction of propiolate (propiolic) with triethylamine (trimethylamine, tertiary amine)' is performed, around 20 references are obtained and one of them (AN 1975:30909) actually is titled 'Reaction of propiolic acid esters with tertiary amines. Formation of betaines'! The point is that this reaction is easily found through **Explore References: Research Topic**, but to find it through any other method would be very challenging.

Second, advantage may be taken of the link to **Get Reactions** (Section 1.2.7) from a bibliographic answer set; i.e. a bibliographic answer set based on a chemical reaction query through **Explore References: Research Topic** may first be obtained and then **Get Reactions** delivers a new answer set in CASREACT. Naturally not all the reactions will be relevant to the initial query, but the analysis and refine options in CASREACT may then be used to narrow reactions.

For example, if information on the Suzuki coupling (reaction of boronic acid derivatives with aryl halides, usually in the presence of palladium catalyst) is required then **Explore References: Research Topic** 'suzuki' may be tried. This gives over 9000 references in CAPLUS and since the **Get Reactions** function is currently restricted to less than 1000 references some further restrictions need to be applied. If the focus is on patent information on heterocyclic boronic acid derivatives then an option is to proceed with **Refine: Document Type** (**Patents**), which affords around 700 references. **Get Reactions** from this answer set then gives around 24,000 reactions in CASREACT. When these are refined with the query shown in Figure 7.21, around 70 patents are retrieved and an example of the Suzuki reaction in one of them is shown in Figure 7.22.

Entry of terms under **Explore References: Research Topic** or **Refine: Research Topic** is also very useful when performed in conjunction with structure or reaction searches. This is illustrated in the section that follows.

## 7.5   Combining Structure, Reaction, Functional Group, and Keyword Terms

The integration of the world's largest chemical substance and bibliographic databases with one of the world's largest chemical reaction databases provides opportunities for searching chemical reaction information in SciFinder that are unique. The user merely

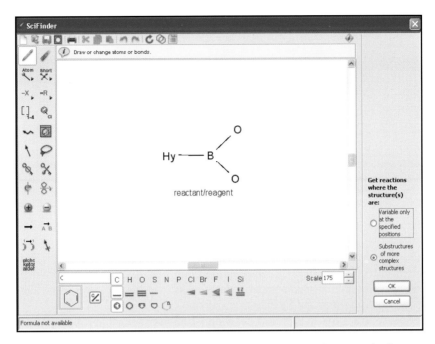

**Figure 7.21** *Query to refine reactions to those which contain heterocyclic boronic acid derivatives. SciFinder® screens are reproduced with permission of Chemical Abstracts Service (CAS), a division of the American Chemical Society*

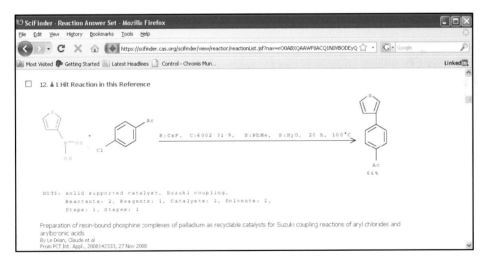

**Figure 7.22** *Sample answer when Suzuki reactions from patents are refined with the query in Figure 7.21. SciFinder® screens are reproduced with permission of Chemical Abstracts Service (CAS), a division of the American Chemical Society*

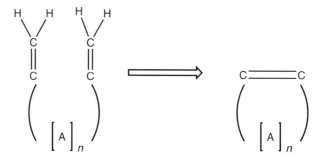

**Figure 7.23**   *General scheme for ring closure metathesis reactions*

has to be aware of the alternatives available and to work through the databases in a creative way.

To illustrate the opportunities, consider the ring closure metathesis reaction pioneered by Professor Robert Grubbs and illustrated in Figure 7.23. Can SciFinder provide some insights into this reaction, perhaps particularly relating to the synthesis of medium size rings (which may or may not include heteroatoms)?

There are several ways to approach this problem. Some thought processes and possible outcomes are:

- **Explore References: Research Topic** 'Grubbs or metathesis', followed by analysis or refine (references), then **Get Reactions** and **Refine: Reactions, Chemical Structure** would be acceptable but at each stage a large number of answers is likely. That is not really a problem because SciFinder post-processing tools can now be used;
- **Explore Substances** (in REGISTRY) would be very difficult, since substance queries would need to be general and issues with system search limits may arise;
- **Explore Reactions** by structure-based queries would exceed system search limits, but a functional group search of ACYCLIC ALKENE (reactant/reagent) and CYCLIC ALKENE (product) would be a reasonable starting point.

When this last option is chosen around 300,000 reactions are retrieved. Since medium size rings are of interest, **Refine** (Reactions) with the query (Figure 7.24) is tried and gives around 1200 reactions, which are reported in around 220 references. The answer in Figure 7.25 is an example of the types of reactions found.

Of these alternatives, the initial functional group explore (around 300,000 reactions) may be considered to afford a reasonable balance between search precision and comprehension, and probably would be the best place to start. The refine option (Figure 7.24) is just one possibility. Among the many others may be to narrow reactions with rings that were of specific sizes, or perhaps first limit to single-step and high yielding reactions. The main difficulty encountered is that many Diels–Alder reactions are retrieved, but they may be eliminated by a number of different structure refinement options. It ultimately depends on the search intent, but by starting with a more general approach the types of issues are encountered and answers may be narrowed in a scientific way.

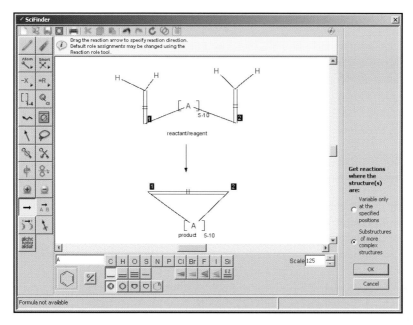

**Figure 7.24** *Initial structure query to explore ring closure metathesis reactions that produce rings of between 7 and 12 atoms. SciFinder® screens are reproduced with permission of Chemical Abstracts Service (CAS), a division of the American Chemical Society*

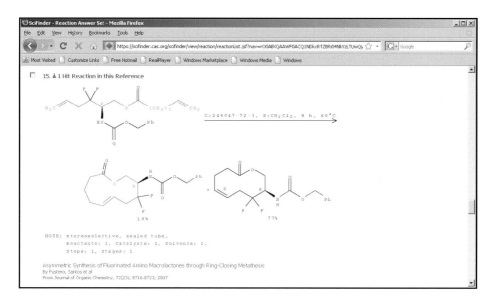

**Figure 7.25** *Sample answer obtained when the query (Figure 7.24) is used to refine initial answers that report the conversion of acyclic alkenes to cyclic alkenes (see text). SciFinder® screens are reproduced with permission of Chemical Abstracts Service (CAS), a division of the American Chemical Society*

At any stage, **Get References** may be used to retrieve bibliographic records and as the search session progresses it is worthwhile checking the indexing of important answers. This is done in the usual way either through looking at key records, or using **Analysis: Index Term**, or using **Categorize**. In the event, it is found that key Index Headings are Metathesis, Metathesis Catalysts, and Macrocyclization, and that the Index Heading Diels–Alder Reaction may be one to use to help eliminate unwanted answers. This may be done through **Refine: Research Topic** 'not diels', although as always caution should be exercised in using the NOT operator.

## 7.6   Summary of Key Points

- Information on chemical reactions may be found in SciFinder through **Explore References: Research Topic** (CAPLUS), through **Explore Substances** (REGISTRY), and through **Explore Reactions** (CASREACT);
- Information from initial answers in one database may easily be transferred to another database:
  - CAPLUS to REGISTRY through **Get Substances**;
  - CAPLUS to CASREACT through **Get Reactions**;
  - CASREACT to CAPLUS through **Get References**;
  - CASREACT to REGISTRY through clicking on a structure in the Reaction Detail screen and then choosing Substance Detail;
  - REGISTRY to CAPLUS through **Get References**, and to find reaction information specifically to choose references in which preparations or reactions are associated with the substance;
  - REGISTRY to CASREACT through **Get Reactions**.
- Once answers in any of the databases are obtained then specific analysis and refinement tools for each database may be used (and **Categorize** may be used in CAPLUS):
  - In particular **Analyze** gives histograms of terms and assists in making decisions on which refinement option to choose.
- CASREACT has a number of special tools to set up search queries, including:
  - Reaction role;
  - Map atoms;
  - Mark bonds (to be formed/broken);
  - Functional group.
- Functional group queries are particularly useful to:
  - Overcome system search limits for general structure queries;
  - Explore selectivity in reactions of functional groups (e.g. include specific additional functional groups to be present in the reaction or specify that additional functional groups are nonreacting).
- Reaction searches may be performed from general to very specific queries:
  - The most specific queries are performed in CASREACT when reactions queries are mapped (atoms/bonds), but generally synthetic chemists are interested in analogous chemistries and so broader queries are usually necessary.

- Retrosynthetic pathways may be investigated through marking bonds to be formed/broken in reaction products:
  - **Analyze: Reactions** by **Product Yield** then helps to determine the most favourable pathways.
- Answers in CASREACT may be:
  - Analysed by bibliographic terms and by reaction information (Catalyst, Number of Steps, Product Yield, and Solvent);
  - Refined by Reaction Structure, Product Yield, Number of Steps, and Reaction Classification.

# Appendix 1
## Some SciFinder Resources

This Appendix contains web links that are referenced in Chapters 1 to 7. The links below are current through June 2009, but changes may occur from time to time as websites develop.

The CAS home page is at www.cas.org and from that site visitors may navigate to a vast amount of information on SciFinder either through:

- Our Expertise, which provides information about the CAS databases and content, or
- Products and Services, which describes the current SciFinder product offering:

  - Support and Training, which offers general training resources as well as solutions to specific types of questions.

Additionally there is a Search (and an Advanced Search) option, which may quickly lead visitors to relevant sites. Table A1.1 below lists some current links for the information indicated.

**Table A1.1**   Some current links for information on SciFinder

| Information available | URL |
| --- | --- |
| General content of CAS databases | www.cas.org/expertise/cascontent/ataglance/index.html |
| Content of CAPLUS | www.cas.org/expertise/cascontent/caplus/index.html |
| Content of REGISTRY | www.cas.org/expertise/cascontent/registry/index.html |
| Content of CASREACT | www.cas.org/expertise/cascontent/casreact.html |
| Content of CHEMCATS | www.cas.org/expertise/cascontent/chemcats.html |
| Content of CHEMLIST | www.cas.org/expertise/cascontent/regulated/index.html |
| Content of MEDLINE | www.nlm.nih.gov/pubs/factsheets/medline.html |
| Patent coverage in CAPLUS | www.cas.org/expertise/cascontent/caplus/patcoverage/worldcov.html |
| Journals in MEDLINE | www.nlm.nih.gov/pubs/factsheets/jsel.html |

*Information Retrieval: SciFinder*®, *Second Edition* Damon D. Ridley
© 2009 John Wiley & Sons, Ltd

**Table A1.1** (*continued*)

| Information available | URL |
|---|---|
| Information on MeSH (MEDLINE subject headings) | www.nlm.nih.gov/pubs/factsheets/mesh.html |
| Information on the MeSH Thesaurus | www.nlm.nih.gov/mesh |
| | www.nlm.nih.gov/mesh/filelist.html |
| MEDLINE allowable qualifiers | www.nlm.nih.gov/mesh/topcat.html |
| SciFinder product information | www.cas.org/products/scifindr/index.html |
| SciFinder support and training | www.cas.org/support/scifi/index.html |
| SciFinder SHelp files | www.cas.org/help/scifinder/index.htm |
| SciFinder analysis function | www.cas.org/help/scifinder/index.htm&ananswr.htm |
| SciFinder refine function | www.cas.org/help/scifinder/index.htm&refine.htm |
| SciFinder categories | www.cas.org/help/scifinder/scicat.htm |
| Description of CA sections | www.cas.org/products/print/ca/casections.html |
| CAS Document Detective Service | www.cas.org/Support/dds.html |

# Appendix 2

# CAS Roles in CAPLUS

CAS Roles are entered after CAS Registry Numbers and Substance Class Headings in CAPLUS. SciFinder offers options to search Roles (e.g. see Figure 3.12) whenever answers are transferred between CAPLUS and REGISTRY. Further options to use CAS Roles are available through histograms from **Analyze: Index Term** and through **Categorize** (particularly Category and Index Terms). Although names of CAS Roles are displayed in CAPLUS records, these names are not searched under **Explore References: Research Topic**.

CAS Roles may be used for certain time periods only; e.g. CAS Role: Combinatorial Study applies in CAPLUS entries from 2002 onwards.

*Information Retrieval: SciFinder®, Second Edition* Damon D. Ridley
© 2009 John Wiley & Sons, Ltd

## CAS roles[1]

### ANST Analytical Study
ANT   Analyte
AMX   Analytical Matrix
ARG   Analytical Reagent Use
ARU   Analytical Role, Unclassified

### BIOL Biological Study
ADV   Adverse Effect, Including Toxicity
AGR   Agricultual Use
BAC   Biological Activity or Effector, Except Adverse (2)
BCP   Biochemical Process (3)
BMF   Bioindustrial Manufacture
BOC   Biological Occurrence (2)
BPN   Biosynthetic Preparation
BPR   Biological Process (2)
BSU   Biological Study, Unclassified
BUU   Biological Use, Unclassified
COS   Cosmetic Use (3)
DGN   Diagnostic Use (3)
DMA   Drug Mechanism of Action (3)
FFD   Food or Feed Use
MFM   Metabolic Formation (2)
NPO   Natual Product Occurrence (3)
PAC   Pharmacological Activity (3)
PKT   Pharmacokinetics (3)
THU   Therapeutic Use

### CMBI Combinatorial Study (3)
CPN   Combinatorial Preparation (3)
CRT   Combinatorial Reactant (3)
CRG   Combinatorial Reagent (3)
CST   Combinatorial Study (3)
CUS   Combinatorial Use (3)

### FORM Formation, Nonpreparative
FMU   Formation, Unclassified
GFM   Geological or Astronomical Formation
MFM   Metabolic Formation (3)

### OCCU Occurrence
BOC   Biological Occurrence (2)
GOC   Geological or Astronomical Occurrence
NPO   Natural Product Occurrence (3)
OCU   Occurrence, Unclassified
POL   Pollutant

### PREP Preparation (4)
BMF   Bioindustrial Manufacture
BPN   Biosynthetic Preparation
BYP   Byproduct
CPN   Combinatorial Preparation (3)
IMF   Industrial Manufacture
PUR   Purification or Recovery
PNU   Preparation, Unclassified (5)
SPN   Synthetic Preparation

### PROC Process
BCP   Biochemical Process (3)
BPR   Biological Process (2)
GPR   Geological or Astronomical Process
PEP   Physical, Engineering, or Chemical Process
    CPS   Chemical Process (6)
    EPR   Engineering Process (6)
    PYP   Physical Process (6)
REM   Removal or Disposal

### PRPH Prophetic Substance (7)

### RACT Reactant or Reagent (2,6)
RCT   Reactant (8)
CRT   Combinatorial Reactant (3)
RGT   Reagent (3)
CRG   Combinatorial Reagent (3)

### USES Uses
AGR   Agricultural use
ARG   Analytical Reagent Use
BUU   Biological Use, Unclassified
CAT   Catalyst Use
COS   Cosmetic Use (3)
CUS   Combinatorial Use (3)
DEV   Device Component Use (5)
DGN   Diagnostic Use (3)
FFD   Food or Feed Use
MOA   Modifier or Additive Use
NUU   Other Use, Unclassified (9)
POF   Polymer in Formulation
TEM   Technical or Engineered Material Use
THU   Therapeutic Use

### Specific roles that are not associated with any super roles:
MSC   Miscellaneous
PRP   Properties

---

(1)  Super roles have 4-letter codes.  Specific roles have 3-letter codes. Under each super role are listed the specific roles that are retrieved when you search that super role.
(2)  Used from CA Vol. 66 (1967) to Vol. 135 (2001).
(3)  Used starting with CA Vol. 136 (2002).
(4)  The PREP super role has been added to records back to 1907.
(5)  Used from CA Vol. 66 (1967) to Vol. 145 (2006).
(6)  Used from CA Vol. 136 (2002) to CA Vol. 145 (2006).

(7)  Used starting with records from CA Vol. 148 (2008).
(8)  Searching the RCT role retrieves references from CA Vol. 66 (1967) to the present.  Searching the RACT super role retrieves references with RCT, CRT, RGT, or CRG references starting with CA Vol. 136 (2002)
(9)  Starting with CA Vol. 136 (2002), the searchable text for the NUU role changed from NONBILOGICAL USE, UNCLASSIFIED/RL to OTHER USE, UNCLASSIFIED/RL. Search NUU/RL to retrieve records from CA Vol. 66 (1967) to the present.

# Appendix 3

## Some Basic Principles Used by SciFinder in the Interpretation of a Research Topic Query

| What SciFinder does | What the implications are | Comments |
|---|---|---|
| **1. CONCEPTS** | | |
| SciFinder uses the prepositions, conjunctions, and stop-words in the question to determine the separate concepts. | Generally it is better to use prepositions between terms and to enter no more than six terms in initial searches. | Try **Explore References: Research Topic** 'measurement *of* mass *of* quarks' rather than 'measurement mass quarks'. Use post-processing tools to narrow answers if needed. |
| If words in the query are not separated by prepositions or conjunctions or stop-words, then SciFinder identifies the words as a 'single' concept in which answers contain the words 'closely associated'. | Searches on exact phrases almost invariably miss important records; 'closely associated' is a better option. *Individual words* in multiword single concepts require *all* the words in the same sentence. | **Explore References: Research Topic** 'treatment of wastewater from gold mining' always keeps the terms 'gold mining' in the same sentence. Try initially 'treatment of wastewater with gold'. |

*Information Retrieval: SciFinder®, Second Edition* Damon D. Ridley
© 2009 John Wiley & Sons, Ltd

| What SciFinder does | What the implications are | Comments |
|---|---|---|
| SciFinder automatically applies truncation and singulars/plurals (except in some cases where the term entered corresponds to an Index Heading or the name of a substance). | This saves considerably on the number of terms needing to be entered, although sometimes the automatic truncation applied may lead to inappropriate hits. If this occurs the solution is to use SciFinder post-processing tools or to select relevant records manually. | **Explore References: Research Topic** 'damon' gives hits on damongo, damonsil, damonia. **Analyze: CA Section Title** allows answers to be narrowed to magnetic and optical studies (e.g. relating to the Damon–Eshbach theory). |
| SciFinder automatically applies synonyms (including common acronyms and CAPLUS/MEDLINE abbreviations) from its synonym dictionary. | This feature saves considerably on effort required to set up the query and gives more comprehensive results; at times, possibly undesired synonyms may be retrieved when the solution is to use SciFinder post-processing tools or to select relevant records manually. | **Explore References: Research Topic** 'sheep' produces hits on lamb(s) and ram(s), but also on RAM. **Analyze: CA Section Title** allows options to remove answers relating to random access memory. |

## 2. CANDIDATES

| | | |
|---|---|---|
| SciFinder presents a list of candidates and first indicates numbers of answers where *all the concepts* are 'closely associated' (usually in the same sentence) and 'anywhere in the reference'. | The concepts identified by SciFinder are presented in bold and in quotes, and first a check should be made to see that the concepts are as intended. Candidates labelled 'closely associated' and 'anywhere in the reference' simply give users alternatives relating to the closeness of the terms (the assumption is the closer the terms, the more directly they are related). | In CAPLUS, an Index Heading is 'closely associated' to its text-modifying phrase. However, different Index Headings are not 'closely associated' (even if they have the same text-modifying phrase). Words in titles and in each sentence in the abstract are 'closely associated'. |

| What SciFinder does | What the implications are | Comments |
|---|---|---|
| SciFinder next indicates the number of answers in which combinations of *some of the concepts* are 'closely associated' or 'anywhere in the reference'. | If three concepts A, B, C are identified, then the number of records for candidates with combinations A and B, A and C, and B and C are listed. | **Explore References: Research Topic** 'the reaction of propiolates with amines' gives around 20 hits where all three concepts are 'closely associated', but almost double the hits with just the two concepts '**propiolates**' and '**amines**' 'closely associated'. This may be a better option. |
| SciFinder next indicates number of answer candidates *for the individual concepts*. | The greater the number of concepts, the greater the restrictions on the answers, and it is helpful to see listings for the individual concepts, particularly when few records are identified with all the concepts present. | **Explore References: Research Topic** 'wastewater from gold mine tailings' shows relatively few hits for the concept '**gold mine tailings**', and the user immediately identifies a potential issue. |
| 3. SYNONYMS | | |
| Users may force inclusion of alternative terms by adding the terms in parentheses. | Care must be taken with 'distributed modifiers' (below). | Use **Explore References: Research Topic** 'chiral reduction (chiral hydrogenation)' rather than 'chiral reduction (hydrogenation)'. |
| 4. BOOLEAN OPERATORS | | |
| SciFinder interprets Boolean AND as a request for both terms *anywhere* in the reference, and the 'closely associated' option is not presented. | Use of AND rather than a preposition does not alert the user to the 'closely associated' option. | It is better to enter **Explore References: Research Topic** 'mass of quarks' rather than 'mass and quarks'. |

| What SciFinder does | What the implications are | Comments |
|---|---|---|
| In some instances, SciFinder may interpret AND as OR. | However, users should enter OR between synonyms. | **Explore References: Research Topic** 'sugars and carbohydrates' gives candidates for OR as well as AND. However, enter 'sugars or carbohydrates'. |
| SciFinder interprets Boolean OR to search for either term. | When entering terms under **Explore References: Research Topic**, it does not matter whether alternative terms are placed in parentheses or linked with OR. | **Explore References: Research Topic** 'steroid analysis in blood of men or women or humans' or 'steroid analysis in blood of men (women, humans)' give similar results. |
| SciFinder interprets Boolean NOT to exclude records with the terms indicated. | The query is interpreted from left to right, so answer sets depend on the placement of the operator. | **Explore References: Research Topic** 'tea and sugar not coffee' is interpreted differently from 'tea not coffee and sugar'. |
| SciFinder does not distribute modifiers. | Words used as modifiers need to be entered with each term to which they refer. | **Explore References: Research Topic** 'chiral reduction (hydrogenation)' identifies concepts **'chiral reduction'** and **'hydrogenation';** therefore enter 'chiral reduction (chiral hydrogenation)'. |

# Appendix 4
## Registration of Substances

**Explore Substances** searches the substance database REGISTRY, which includes all types of chemical substances such as organic, inorganic, alloys, polymers, mixtures, reactive intermediates, proteins, nucleic acids, and so forth. In most cases, the registration of substances follows exactly the valence bond descriptions taught in chemistry courses. However, the many subtle variations in structures often require modification of valence bond descriptions and the key issues are resonance, tautomerism, $\pi$-complexes, $\sigma$-complexes, radicals, and other reactive intermediates. Specific rules are applied when structures with these variations are entered into computer databases.

Computer databases also need to have rules for defining, among other things, salts, mixtures, hydrates, polymers, and alloys in which valence bond descriptions do not necessarily apply. Many of these are addressed by registering the substance as one made up of a number of components. About 10% of substances in REGISTRY are multicomponent substances, and the key is to recognize that each component is identified as a single entry in the formula and in the structure fields.

A good way to learn about the registration of substances is to examine actual records and this appendix gives examples. If further information is required it is suggested that users retrieve the substances in this appendix in SciFinder (e.g. by entering the CAS Registry Numbers of the substances in **Explore Substance**s: **Substance Identifier**).

## A4.1   Single-Component Substances

### A4.1.1   Single Substances

The most common class is that of single substances; typical records are shown for cortisone (Figure A4.1) and epibatidine (Figure A4.2)[1].

#### A4.1.1.1   Notes

1. When the name of a substance is entered through **Explore Substances: Substance Identifier**, SciFinder initially seeks a match with a name in REGISTRY. If there is

---

[1] SciFinder screens in this appendix are reproduced with permission of Chemical Abstracts Service (CAS), a division of the American Chemical Society.

*Information Retrieval: SciFinder®, Second Edition* Damon D. Ridley
© 2009 John Wiley & Sons, Ltd

CAS Registry Number: 53-06-5

C₂₁ H₂₈ O₅

Pregn-4-ene-3,11,20-trione, 17,21-dihydroxy-

Cortisone (8CI) ; 11-Dehydro-17-hydroxycorticosterone ; 17,21-Dihydroxypregn-4-ene-3,11,20-trione ; 17-Hydroxy-11-dehydrocorticosterone ; 17α,21-Dihydroxy-4-pregnene-3,11,20-trione ; 17α,21β-Dihydroxy-4-pregnene-3,11,20-trione ; 17α-Hydroxy-11-dehydrocorticosterone ; Adrenalex ; Compound E ; Cortisate ; Cortivite ; Cortogen ; Cortone ; KE ; Kendall's compound E ; NSC 9703 ; Pregn-4-en-17α,21-diol-3,11,20-trione ; Reichstein's substance Fa ; Wintersteiner's compound F ; Δ4-Pregnene-17α,21-diol-3,11,20-trione

Component

Deleted CAS Registry Numbers: 478614-18-5

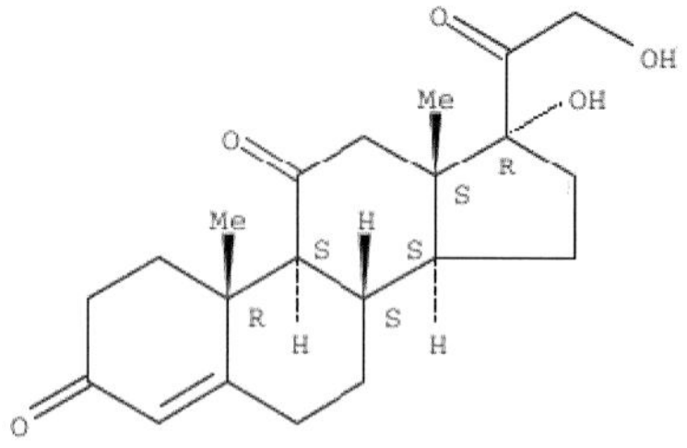

Absolute stereochemistry.

*Figure A4.1*

CAS Registry Number: 140111-52-0

C₁₁ H₁₃ Cl N₂

7-Azabicyclo[2.2.1]heptane, 2-(6-chloro-3-pyridinyl)-, (1R,2R,4S)-

7-Azabicyclo[2.2.1]heptane, 2-(6-chloro-3-pyridinyl)-, (1R-exo)- ; (+)-Epibatidine ; (1R,4S)-Epibatidine ; (1R-exo)-2-(6-Chloro-3-pyridinyl)-7-azabicyclo[2.2.1]heptane ; Alkaloid 208/210 from Dendrobates ; CMI 488 ; Epibatidine

Component

Deleted CAS Registry Numbers: 152378-31-9; 163437-11-4

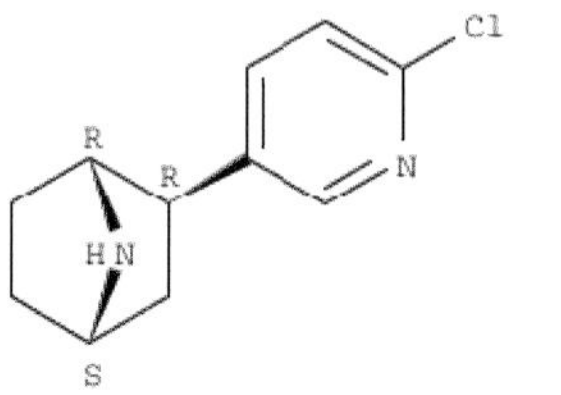

Absolute stereochemistry. Rotation (+).

*Figure A4.2*

not an exact match, then SciFinder attempts to find possible answers on the basis of parts of names (Section 4.5).

2. When terms are entered under **Explore References: Research Topic**, SciFinder checks to see whether any of the terms exactly matches a name in REGISTRY. If there is an exact match, the CAS Registry Number is included in the 'concept' searched.

3. Additional search options (Figure 4.3) include an option to search for single-component substances only.

### A4.1.2    Elements, Ions, and Particles

All elements, ions, and subatomic particles are registered, and examples are given in Figure A4.3 (calcium), Figure A4.4 (calcium 2⁺), and Figure A4.5 (strange quark).

*A4.1.2.1    Notes*

1. CAS Registry Numbers for elements are indexed when the 'element' is mentioned in the original record (e.g. CAS Registry Numbers for silicon and for calcium

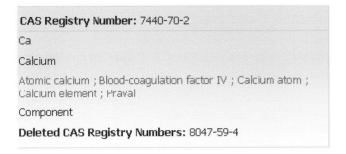

*Figure A4.3*

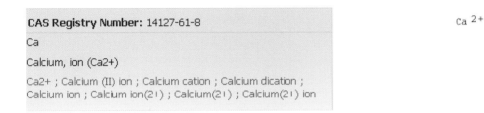

*Figure A4.4*

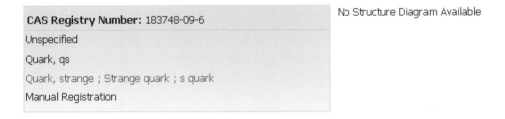

*Figure A4.5*

are entered when 'silicon wafers' and 'blood calcium levels' respectively are mentioned).

2. CAS Registry Numbers for ions are indexed when the ions are mentioned in the original record (e.g. 'calcium ion levels in blood').

3. All subatomic particles have CAS Registry Numbers that are the preferred search terms – for precision and comprehension.

### A4.1.3 Isotopic Substances

Isotopes of hydrogen are indicated by the symbols D and T (Figure A4.6), while isotopes of other atoms have the atomic weight as a superscript before the symbol for the atom (Figure A4.7). These representations appear in the structure diagram, in the name entries, and, in the case of hydrogen isotopes, in the formula.

CAS Registry Number: 105089-96-1

C₁₀ H₁₃ D N₂

Pyridine, 3-(1-methyl-2-pyrrolidinyl-5-d)-, (2S-cis)- (9CI)
Component

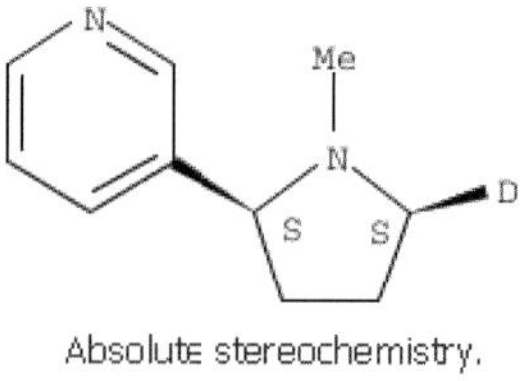

Absolute stereochemistry.

Source of Registration: CA

**Figure A4.6**

CAS Registry Number: 67209-85-2

C₁₀ H₁₄ N₂

Pyridine, 3-[1-(methyl-13C)-2-pyrrolidinyl-2,3-13C2]- (9CI)
Pyridine, 3-[1-(methyl-13C)-2-pyrrolidinyl-2,3-13C2]-, (±)-
Component

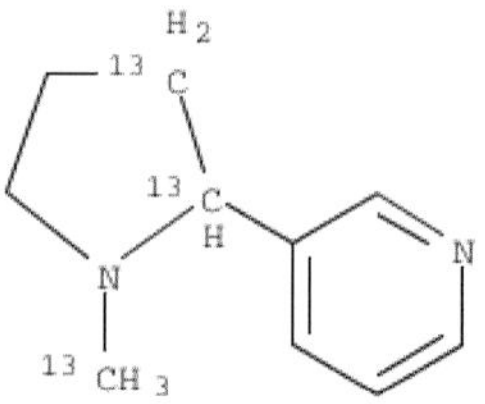

**Figure A4.7**

### A4.1.3.1   Notes

1. Since isotopes of hydrogen (D, T) are included in the molecular formula, **Explore Substances: Molecular Formula** may be used to find substances with these isotopes.
2. Other isotopic substances are found by searching the structure (exact search) and then scrolling through the answers. **Refine: Isotope-Containing** (Section 5.3.2) may also be used to narrow answers to less common isotopes.

### A4.1.4   Stereoisomers

The letters R, S, E, and Z and the Greek letters $\alpha$ and $\beta$ are commonly used to indicate stereochemical arrangements in molecules. Structure diagrams additionally indicate stereochemistry through heavy, dotted, or wedged bonds.

The registration of an enantiomer and of a racemic form are shown in Figures A4.8 and A4.9. Note the stereochemical descriptors ((S), L, and (+); (R,S), DL, dl, and (.+ − .)) in the name fields.

Cortisone (Figure A4.1), epibatidine (Figure A4.2), augmentin (Figure A4.10), and nicotine hydroiodide (Figure A4.11) have stereochemical features, and the ways in which these are presented in the name and structure fields should be noted.

Geometrical isomers receive separate registrations. The substances with stereochemistry undefined and the (Z)-isomer are shown in Figures A4.12 and A4.13.

### A4.1.4.1   Notes

1. Substances are entered as precisely as possible. If the stereochemistry is specified in the original document, then the specific stereoisomer is indexed.

CAS Registry Number: 56-41-7

$C_3 H_7 N O_2$

L-Alanine

Alanine, L- (7CI,8CI) ; (2S)-2-Aminopropanoic acid ; (S)-(+)-Alanine ; (S)-2-Aminopropanoic acid ; (S)-Alanine ; 1561: PN: US20060223088 SEQID: 1571 claimed protein ; 1634: PN: US20070015696 SEQID: 1644 claimed protein ; Alanine ; L-(+)-Alanine ; L-2-Aminopropanoic acid ; L-2-Aminopropionic acid ; L-α-Alanine ; L-α-Aminopropionic acid ; Lactamine ; NSC 206315 ; Propanoic acid, 2-amino-, (S)- ; α-Alanine ; α-Aminopropionic acid

Component

Deleted CAS Registry Numbers: 6898-94-8; 115967-49-2; 170805-71-7; 759445-89-1; 787635-21-6

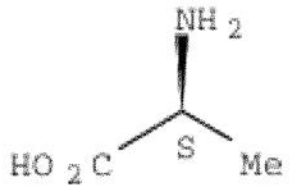

Absolute stereochemistry. Rotation (+).

**Figure A4.8**

CAS Registry Number: 302-72-7

$C_3 H_7 N O_2$

Alanine

Alanine, DL- (8CI) ; DL-Alanine ; (R,S)-Alanine ; (±)-2-Aminopropionic acid ; (±)-Alanine ; DL-Ala ; DL-α-Alanine ; DL-α-Aminopropionic acid ; NSC 7602 ; dl-2-Aminopropanoic acid ; dl-Alanine

Component

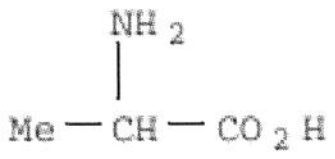

**Figure A4.9**

2. The simplest way to find stereoisomers is to do a structure search (exact or substructure) and to look through the answers.

## A4.1.5 Donor Bonds

When one of the atoms in the bond provides both of the bond electrons, the structure is represented with a double bond. This occurs in particular with higher oxidation states of metals and of nonmetals such as nitrogen, phosphorus, and sulfur. In some instances, this representation will produce a 'structure' that violates valence bond rules (e.g. the valence of 4 for nitrogen in 4-nitropyridine 1-oxide; see Figure A4.14).

### A4.1.5.1 Notes

1. In structure searches SciFinder automatically recognizes compounds with donor bonds and searches the appropriate structure. For example, the substance (Figure A4.14) is retrieved irrespective of whether a double or a single bond is drawn between the nitrogen and the oxygen.
2. The shortcut symbol for the nitro group is used in structure displays, but if drawn in full there are 'double' bonds from the nitrogen to each of the oxygens.

CAS Registry Number: 74469-00-4

$C_{16} H_{19} N_3 O_5 S . C_8 H_9 N O_5 . K$

4-Oxa-1-azabicyclo[3.2.0]heptane-2-carboxylic acid, 3-(2-hydroxyethylidene)-7-oxo-, potassium salt (1:1), (2R,3Z,5R)-, mixt. with (2S,5R,6R)-6-[[(2R)-2-amino-2-(4-hydroxyphenyl)acetyl]amino]-3,3-dimethyl-7-oxo-4-thia-1-azabicyclo[3.2.0]heptane-2-carboxylic acid

4-Oxa-1-azabicyclo[3.2.0]heptane-2-carboxylic acid, 3-(2-hydroxyethylidene)-7-oxo-, monopotassium salt, (2R,3Z,5R)-, mixt. with (2S,5R,6R)-6-[[(2R)-amino(4-hydroxyphenyl)acetyl]amino]-3,3-dimethyl-7-oxo-4-thia-1-azabicyclo[3.2.0]heptane-2-carboxylic acid (9CI) ; 4-Oxa-1-azabicyclo[3.2.0]heptane-2-carboxylic acid, 3-(2-hydroxyethylidene)-7-oxo-, monopotassium salt, [2R-(2α,3Z,5α)]-, mixt. with [2S-[2α,5α,6β(S*)]]-6-[[amino(4-hydroxyphenyl)acetyl]amino]-3,3-dimethyl-7-oxo-4-thia-1-azabicyclo[3.2.0]heptane-2-carboxylic acid ; 4-Thia-1-azabicyclo[3.2.0]heptane-2-carboxylic acid, 6-[[(2R)-amino(4-hydroxyphenyl)acetyl]amino]-3,3-dimethyl-7-oxo-, (2S,5R,6R)-, mixt. contg. (9CI) ; 4-Thia-1-azabicyclo[3.2.0]heptane-2-carboxylic acid, 6-[[amino(4-hydroxyphenyl)acetyl]amino]-3,3-dimethyl-7-oxo-, [2S-[2α,5α,6β(S*)]]-, mixt. contg. ; Amocla ; Amoclan ; Amoclav ; Amoran ; Amoxicillin-potassium clavulanate mixt. ; Amoxsiklav ; Ancla ; Augmentan ; Augmentin ; Augmentin (antibiotic) ; Augmentin ES 600 ; Augmentine ; Auspilic ; BRL 25000 ; BRL 25000A ; BRL 25000G ; Clamentin ; Clamobit ; Clamonex ; Clamoxyl ; Clavamox ; Clavinex ; Clavoxilin Plus ; Clavulin ; Clavumox ; Eumetinex ; Forcid Solutab ; Kmoxilin ; Spectramox ; Viaclav ; Xiclav ; co-Amoxyclav

Mixture

**Deleted CAS Registry Numbers:** 74428-36-7

61177-45-5 (Component: 58001-44-8)

$C_8 H_9 N O_5 . K$

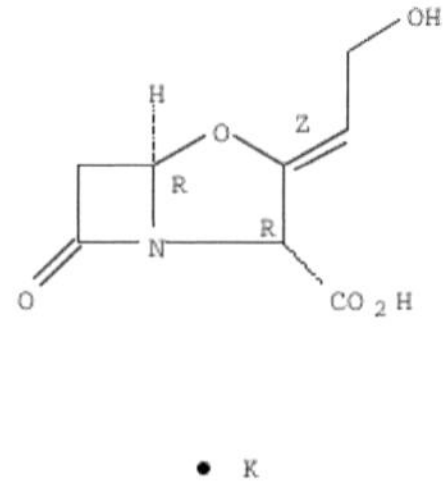

• K

Absolute stereochemistry.
Double bond geometry as shown.

26787-78-0

$C_{16} H_{19} N_3 O_5 S$

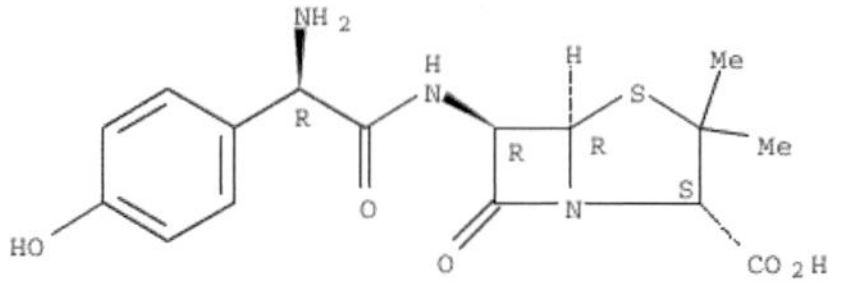

Absolute stereochemistry.

*Figure A4.10*

CAS Registry Number: 6019-03-0
(Component: 54-11-5)

$C_{10} H_{14} N_2 . H I$

Pyridine, 3-[(2S)-1-methyl-2-pyrrolidinyl] , hydriodide (1:1)

Nicotine, monohydriodide (8CI) ; Pyridine, 3-(1-methyl-2-pyrrolidinyl)-, monohydriodide, (S)- (9CI) ; Nicotine hydriodide ; Nicotine hydroiodide

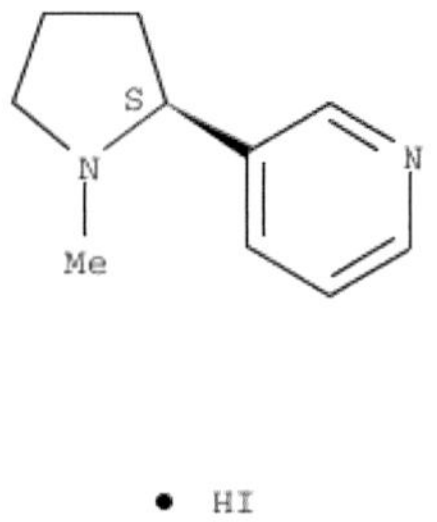

• HI

Absolute stereochemistry. Rotation (-).

*Figure A4.11*

## A4.1.6  Intermediates

Examples of a carbene and a radical are shown in Figures A4.15 and A4.16 respectively. However, the CAS Registry Number for reactive intermediates will be inserted in CAPLUS only when the intermediates are clearly identified or are an important part of the original paper. Intermediates drawn in mechanistic schemes are not indexed (unless they satisfy the identification criteria).

### A4.1.6.1  Notes

1. Most radicals, carbocations, and carbanions have relatively unusual formulas (because of the odd valency at carbon), so **Explore Substances: Molecular Formula** usually

**CAS Registry Number:** 104-98-3

$C_6 H_6 N_2 O_2$

2-Propenoic acid, 3-(1H-imidazol-5-yl)-

2-Propenoic acid, 3-(1H-imidazol-4-yl)- (9CI) ; Imidazole-4-acrylic acid (8CI) ; 3-(1H-Imidazol-4-yl)-2-propenoic acid ; 3-(1H-Imidazol-4-yl)acrylic acid ; 3-(4-Imidazolyl)acrylic acid ; 5-Imidazoleacrylic acid ; NSC 66357 ; Urocanic acid ; Urocaninic acid

Component

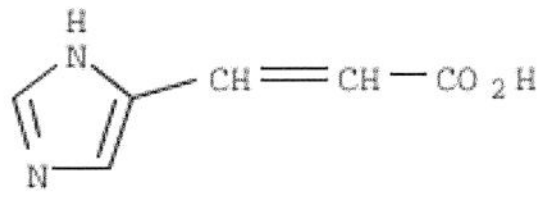

*Figure A4.12*

**CAS Registry Number:** 7699-35-6

$C_6 H_6 N_2 O_2$

2-Propenoic acid, 3-(1H-imidazol-5-yl)-, (2Z)-

2-Propenoic acid, 3-(1H-imidazol-4-yl)-, (2Z)- (9CI) ; 2-Propenoic acid, 3-(1H-imidazol-4-yl)-. (Z)- ; Imidazole-4-acrylic acid, (Z)- (8CI) ; (Z)-Urocanic acid ; cis-Urocanic acid

Component

**Deleted CAS Registry Numbers:** 738555-25-4; 927524-97-8

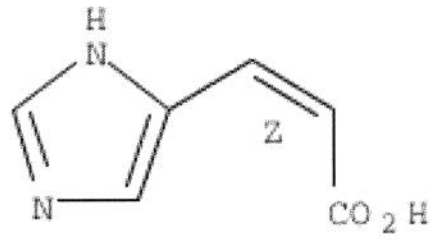

Double bond geometry as shown.

*Figure A4.13*

**CAS Registry Number:** 1124-33-0

$C_5 H_4 N_2 O_3$

Pyridine, 4-nitro-, 1-oxide

4-Nitropyridine 1-oxide ; 4-Nitropyridine N-oxide ; 4-Nitropyridine oxide ; NSC 130895 ; NSC 5079 ; p-Nitropyridine N-oxide

Component

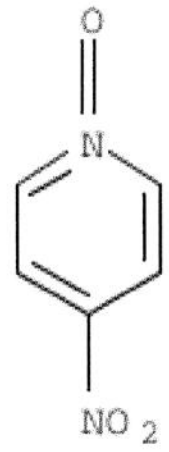

*Figure A4.14*

**CAS Registry Number:** 39922-09-3

$C_6 H_{10}$

Cyclohexylidene

Carbenacyclohexane

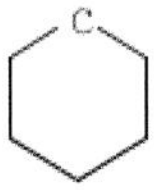

*Figure A4.15*

CAS Registry Number: 67271-34-5

C₇ H₁₃

Methyl, cyclohexyl- (9CI)

Cyclohexylmethyl radical

*Figure A4.16*

retrieves the intermediates quickly. If needed, **Refine: Chemical Structure** followed by drawing, then searching, the required carbon skeleton may further narrow answer sets (Section 4.6.1).

2. Carbenes are isomeric with alkenes, so initial searches on molecular formulas will produce larger answer sets that need to be refined (e.g. by structure).

3. Because of the odd valency of carbon in all these intermediates, substructure searches on the ring skeletons may not retrieve intermediates, so an initial molecular formula search may be necessary.

## A4.2    Multicomponent Substances

A record for a multicomponent substance lists only a few of the names for the individual substances involved, has a 'dot disconnected' entry in the formula field, and has the individual components presented in the structure display (which also gives the CAS Registry Number of the component). Multicomponent substances are commonly encountered in copolymers, salts, mixtures, minerals, and alloys. At present, in the substance database, there are more than 6 million substances with two or more components.

Note that an entry 'Component' appears after the substance name(s) when the CAS Registry Number appears as a component in a multicomponent substance and this alerts the user to consider related substances. For example, the record for epibatidine (Figure A4.2) contains the entry 'Component' and the user is alerted to the fact that epibatidine appears in multicomponent substances. In this case the multicomponent substances are mainly salts (e.g. the hydrochloride) of epibatidine and the need to search for such salts should be considered.

### A4.2.1    Salts

*A4.2.1.1    Salts That Do Not Contain Carbon*

Simple salts from acids not containing Periodic Table Group VI atoms (e.g. oxygen and sulfur) are registered in the way normally drawn by the chemist. For example, sodium chloride and calcium bromide are represented as NaCl and CaBr₂ respectively.

However, salts from acids containing Periodic Table Group VI atoms and bases containing Periodic Table Group I and II atoms (e.g. sodium, potassium, calcium, barium) are registered as the free acid combined with the base (which is given the metal symbol). For example, although chemists write the formula for the calcium phosphate as $Ca_3(PO_4)_2$ the substance is registered as if the hydrogens were still attached to the phosphate group, i.e. as phosphoric acid, $H_3PO_4$. As the actual salt has three Ca atoms

**CAS Registry Number:** 7758-87-4
(Component: 7664-38-2)

Ca . $^2/_3$ H$_3$ O$_4$ P

Phosphoric acid, calcium salt (2:3)

Allogran R ; Aparmicron AP 12C ; Bonarka ; C 13 09 ; C 13 09SF ; Calcium orthophosphate ; Calcium orthophosphate (Ca3(PO4)2) ; Calcium phosphate ; Calcium phosphate (3:2) ; Calcium phosphate (Ca3(PO4)2) ; Calcium phosphate Ca0.75(PO4)0.5 ; Calcium tertiary phosphate ; Calipharm T ; Cerasorb ; Ceredex ; Coveg ; DET 10 ; Multifos ; Osferion ; Osferion G1 ; Ossaplast ; Ostram ; Phos Calcium P 24 ; Phosphoric acid calcium(2+) salt (2:3) ; Posture ; Posture (calcium supplement) ; Synthcgraft ; Synthos ; TCP ; TCP 10 ; TCP 10U ; TCP 118FG ; Tertiary calcium phosphate ; Tribasic calcium phosphate ; TricOs ; Tricafos P ; Tricalcium diphosphate ; Tricalcium orthophosphate ; Tricalcium phosphate ; Tricalcium phosphate (Ca3(PO4)2) ; Vitoss ; α-Tricalcium phosphate ; β-TCP ; β-Tricalcium phosphate ; β-Whitlockite

Component

**Deleted CAS Registry Numbers:** 1344-15-6; 123211-19-8; 915772-79-1

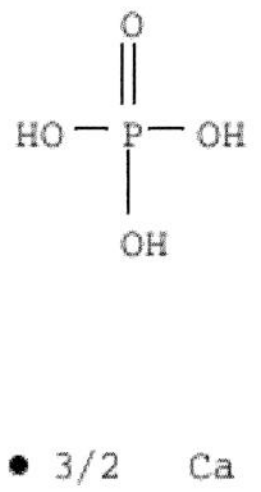

$$\begin{array}{c} O \\ \| \\ HO - P - OH \\ | \\ OH \end{array}$$

● 3/2    Ca

*Figure A4.17*

to two phosphate groups, the substance is considered as $Ca_3(H_3PO_4)_2$, and the molecular formula, which has the first component in the alphabet as a single atom, then becomes Ca. 2/3 H$_3$ O$_4$ P (Figure A4.17).

Actually, there are two substances called *calcium phosphate* in the database and only one is shown here (the Molecular Formula Field entry for the second substance is Ca. x H$_3$ O$_4$ P). The issue is the chemical description of the substances in the original literature. If the particular form is not specified in the article, the substance is registered as that in which the ratio is not specified. There are currently more than 10,000 entries for this second substance, so a search for information on calcium phosphate should probably include its CAS Registry Number (10103-46-5) as well.

### A4.2.1.2    Salts from Organic Acids and Periodic Table Group I and II Bases

The registration is similar to the registration described above. Note that the molecular formula for sodium acetate is listed as C$_2$ H$_4$ O. Na (Figure A4.18), whereas a chemist would represent the formula for the substance as $C_2H_3ONa$.

### A4.2.1.3    Salts with Nitrogen-Containing Bases

When the base is an amine the salts are represented as the free base and the free acid. An example is shown in Figure A4.11.

Salts from nitrogen-containing bases involving other acids (e.g. sulfuric acid and phosphoric acid) are registered similarly (e.g. ammonium sulfate is the multicomponent substance with the molecular formula H$_3$N. $^1/_2$ H$_2$ O$_4$ S) although the molecular formula

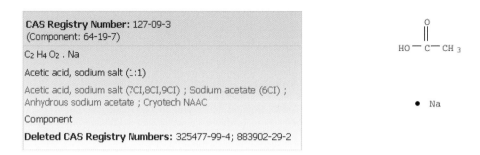

**CAS Registry Number:** 127-09-3
(Component: 64-19-7)

$C_2 H_4 O_2$ . Na

Acetic acid, sodium salt (1:1)

Acetic acid, sodium salt (7CI,8CI,9CI) ; Sodium acetate (6CI) ;
Anhydrous sodium acetate ; Cryotech NAAC

Component

**Deleted CAS Registry Numbers:** 325477-99-4; 883902-29-2

*Figure A4.18*

for ammonium chloride is Cl $H_4$ N. These substances are most easily found through searches based on names.

### A4.2.1.4    Salts from Organic Acids and Organic Bases

These salts are registered as the free acid and the free base, but the subtle difference is that now both the acid and the base are registered as individual components in the two-component registration (Figure A4.19).

### A4.2.1.5    Notes

1. Many biologically important substances are salts and have different CAS Registry Numbers from the parent acid or base, so it is important to consider search strategies that will retrieve the parent substance including all its salts.
2. Salts are retrieved through an exact structure search on either the parent acid or base.
3. To retrieve specific salts, draw both structure fragments on the same screen and then perform an exact structure search.

## A4.2.2    Alloys

When specified in the original article, the composition of the alloy is listed in the name and composition fields and the constituent elements are listed in the formula field

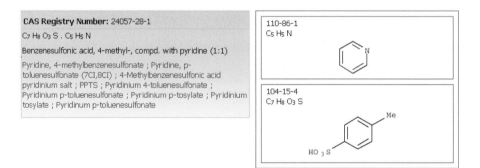

**CAS Registry Number:** 24057-28-1

$C_7 H_8 O_3 S$ . $C_5 H_5 N$

Benzenesulfonic acid, 4-methyl-, compd. with pyridine (1:1)

Pyridine, 4-methylbenzenesulfonate ; Pyridine, p-
toluenesulfonate (7CI,8CI) ; 4-Methylbenzenesulfonic acid
pyridinium salt ; PPTS ; Pyridinium 4-toluenesulfonate ;
Pyridinium p-toluenesulfonate ; Pyridinium p-tosylate ; Pyridinium
tosylate ; Pyridinum p-toluenesulfonate

110-86-1

$C_5 H_5 N$

104-15-4

$C_7 H_8 O_3 S$

*Figure A4.19*

**CAS Registry Number:** 11109-50-5

C . Cr . Fe . Mn . Ni . P . S . Si

Iron alloy, base, Fe 66-74,Cr 18.00-20.00,Ni 8.00-10.50,Mn 0-2.00,Si 0-1.00,C 0-0.08,P 0-0.045,S 0-0.030 (UNS S30400)

Chromium alloy, nonbase, Fe 66-74,Cr 18.00-20.00,Ni 8.00-10.50,Mn 0-2.00,Si 0-1.00,C 0-0.08,P 0-0.015,S 0-0.030 (UNS S30400) ; Manganese alloy, nonbase, Fe 66-74,Cr 18.00-20.00, Ni 8.00-10.50,Mn 0-2.00,Si 0-1.00,C 0-0.08,P 0-0.045,S 0-0.030 (UNS S30400) ; Nickel alloy, nonbase, Fe 66-74,Cr 18.00-20.00, Ni 8.00-10.50,Mn 0-2.00,Si 0-1.00,C 0-0.08,P 0-0.045,S 0-0.030 (UNS S30400) ; Silicon alloy, nonbase, Fe 66-74,Cr 18.00-20.00, Ni 8.00-10.50,Mn 0-2.00,Si 0-1.00,C 0-0.08,P 0-0.045,S 0-0.030 (UNS S30400) ; 0.09C18Cr9Ni ; 06Cr18Ni10 ; 06Kh18N10 ; 07Cr18Ni10 ; 08Kh18N10 ; 08Kh18N9 ; 09Kh18N9 ; 0Cr18Ni9 ;

Alloy

**Deleted CAS Registry Numbers:** 12618-33-6; 12653-18-8; 12718-13-7; 12743-59-8; 12744-08-0; 37195-28-1; 37301-63-6; 37329-54-7; 39424-45-8; 50812-62-9; 51204-61-6; 52013-09-9; 55919-24-9; 55964-80-2; 57175-33-4; 67661-38-5; 68856-45-1; 70689-12-2; 85799-92-4; 88705-07-1; 88705-08-2; 101845-49-2; 140608-02-2; 146633-47-8; 147129-89-3; 154763-95-8; 186845-06-7; 258821-65-7; 433921-75-6; 780038-69-9; 957464-24-3; 1013331-25-3

Composition

| Component | Component Percent | Component Registry Number |
|---|---|---|
| Fe | 66 - 74 | 7439-89-6 |
| Cr | 18.00 - 20.00 | 7440-47-3 |
| Ni | 8.00 - 10.50 | 7440-02-0 |
| Mn | 0 - 2.00 | 7439-96-5 |
| Si | 0 - 1.00 | 7440-21-3 |
| C | 0 - 0.08 | 7440-44-0 |
| P | 0 - 0.045 | 7723-14-0 |
| S | 0 - 0.030 | 7704-34-9 |

*Figure A4.20*

No Structure Diagram Available

**CAS Registry Number:** 12597-68-1

Unspecified

Stainless steel

Austenitic stainless steel ; Chromium nickel stainless steel ; Chromium stainless steel ; Nickel stainless steel ; Pall Ring ; STN 350

Alloy Manual Registration

**Deleted CAS Registry Numbers:** 68445-95-4; 179466-28-5; 179466-32-1

*Figure A4.21*

(Figure A4.20). A more generic description (Figure A4.21) is applied when the elemental composition is not precisely specified in the original article.

*A4.2.2.1   Notes*

1. Most alloys are easily retrieved through molecular formula searches, and if special compositions are required, it is necessary to look through individual records.
2. Alloys may also be retrieved by drawing the separate elements in the structure screen and then using exact or substructure search options as needed.

**CAS Registry Number:** 67713-16-0

$C_7 H_5 N_3 O_6 . C_3 H_6 N_6 O_6 . Al$

1,3,5-Triazine, hexahydro-1,3,5-trinitro-, mixt. with aluminum and 2-methyl-1,3,5-trinitrobenzene

Aluminum, mixt. contg. (9CI) ; Benzene, 2-methyl-1,3,5-trinitro-, mixt. contg. (9CI) ; Alex 20 ; Alex 32 ; Comp B ; Composition A ; Composition B ; Composition C ; Composition D ; H 6 ; HBX 1 ; HBX 3 ; Hexotonal ; KS 22 ; Torpex

Mixture

**Deleted CAS Registry Numbers:** 57608-65-8; 57608-66-9; 60976-68-3; 78403-24-4

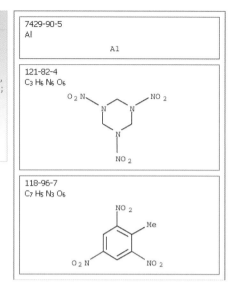

7429-90-5

Al

A1

121-82-4

$C_3 H_6 N_6 O_6$

118-96-7

$C_7 H_5 N_3 O_6$

*Figure A4.22*

### A4.2.3  Mixtures

Mixtures are registered where two or more chemically discrete components have been mixed together for a specific use (e.g. formulations involving pharmaceutical and agricultural chemicals). In general, host–guest complexes are also considered as mixtures (see Figures A4.10 and A4.22).

*A4.2.3.1  Notes*

1. An exact structure search will retrieve substances in which the substance is a component of a mixture.
2. To search for specific multicomponent substances, draw the separate components on the structure editor and search 'exact'.

## A4.3  Metal Complexes

The representation of the structures of some coordination compounds requires modifications to normal valence bond definitions. Generally, the electrons involved in bonding the organic groups to the metal are provided by the organic groups as either $\sigma$-donors and $\pi$-donors. In broad terms, these are distinguished in that the electrons in the former case come from atoms, whereas in the latter case they come from double bonds.

Another issue is that the difference between a coordination compound and a salt may be difficult to define (see Figures A4.23 and A4.24). The following examples are illustrative.

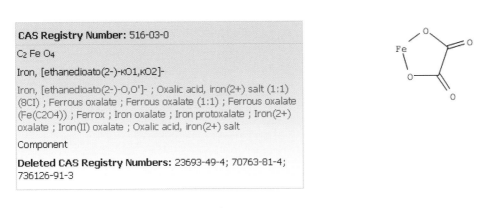

CAS Registry Number: 3094-87-9
(Component: 64-19-7)

$C_2 H_4 O_2 . \frac{1}{2} Fe$

Acetic acid, iron(2+) salt (2:1)

Acetic acid, iron(2+) salt (8CI,9CI) ; Ferrous acetate ; Iron acetate [Fe(OAc)2] ; Iron diacetate ; Iron(2+) acetate ; Iron(II) acetate

Component

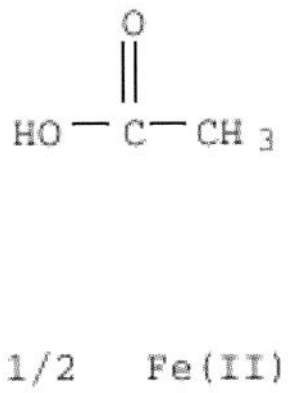

*Figure A4.23*

CAS Registry Number: 516-03-0

$C_2 Fe O_4$

Iron, [ethanedioato(2-)-κO1,κO2]-

Iron, [ethanedioato(2-)-O,O']- ; Oxalic acid, iron(2+) salt (1:1) (8CI) ; Ferrous oxalate ; Ferrous oxalate (1:1) ; Ferrous oxalate (Fe(C2O4)) ; Ferrox ; Iron oxalate ; Iron protoxalate ; Iron(2+) oxalate ; Iron(II) oxalate ; Oxalic acid, iron(2+) salt

Component

Deleted CAS Registry Numbers: 23693-49-4; 70763-81-4; 736126-91-3

*Figure A4.24*

## A4.3.1   σ-Complexes

Generally, charges in structures relate to the species involved in the preparation of the complex. For example, cisplatin (Figure A4.25) is made from $Pt^{2+}$, $Cl^-$, and $NH_3$.

## A4.3.2   π-Complexes

In π-complexes an extra 'bond' is drawn between the atoms involved in the complex (Figure A4.26).

### A4.3.2.1   Notes

1. Bonds to metals in queries may be ignored in the initial structure search. In these cases, **Show precision analysis** may be required to obtain more precise answer sets.
2. It is not necessary to insert charges in structure queries.
3. This representation of π-complexes may produce structures where the normal valencies of atoms are exceeded (e.g. ferrocenes derived from pentamethylcyclopentadiene), but structure searches proceed as expected.

**CAS Registry Number:** 15663-27-1

Cl₂ H₆ N₂ Pt

Platinum, diamminedichloro-, (SP-4-2)-

Platinum, diamminedichloro-, cis- (8CI) ; Abiplatin ; Biocisplatinum ; Briplatin ; CACP ; CDDP ; CPDC ; CPDD ; CPPD ; Cismaplat ; Cisplatin ; Cisplatino ; Cisplatinum ; Cisplatyl ; Citoplatino ; DDP ; DDP (antitumor agent) ; Fauldiscipla ; Lederplatin ; Lipoplatin ; NSC 119875 ; Neoplatin ; Platamine ; Platiblastin ; Platidiam ; Platinex ; Platinol ; Platinol AQ ; Platinoxan ; Platistin ; Platosin ; Rand ; Randa ; SPI 077 ; SPI 077 (complex) ; SPI 077B103 ; TR 170 ; cis-DDP ; cis-Diaminedichloroplatinum(II) ; cis-Diaminodichloroplatinum(II) ; cis-Diamminedichloroplatinum ; cis-Diamminedichloroplatinum(II) ; cis-Dichlorodiamineplatinum(II) ; cis-Dichlorodiammineplatinum ; cis-Dichlorodiammineplatinum(II) ; cis-Platin ; cis-Platine ; cis-Platinous diaminodichloride ; cis-Platinum ; cis-Platinum II ; cis-Platinum diaminodichloride ; cis-Platinum(II) diaminodichloride ; cis-Platinum(II) diamminedichloride ; cis-Platinumdiamine dichloride ; cis-Platinumdiammine dichloride

Coordination Compound Component

**Deleted CAS Registry Numbers:** 96081-74-2; 936542-99-3

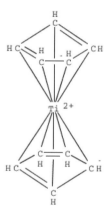

*Figure A4.25*

**CAS Registry Number:** 1271-29-0

C₁₀ H₁₀ Ti

Titanocene

Titanium, di-n-cyclopentadienyl- (8CI) ; Titanium, dicyclopentadienyl- (6CI,7CI) ; Bis(η5-cyclopentadienyl)titanium ; Dicyclopentadienyltitanium

Ring Parent Coordination Compound Component

**Deleted CAS Registry Numbers:** 11084-64-3; 11137-38-5

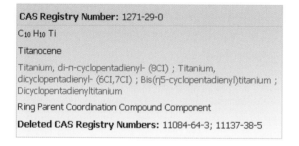

*Figure A4.26*

## A4.4   Macromolecules

The primary registration of polymers is as their monomer components, which means that copolymers are registered as multicomponent substances (Section A4.4.2). In cases where the chemistry involved in the polymerization means the polymer must have a single-structure repeating unit, then a supplementary registration is applied (Section A4.4.3).   However, some polymers of this type are indexed only as the structure repeating unit (Section 6.10).

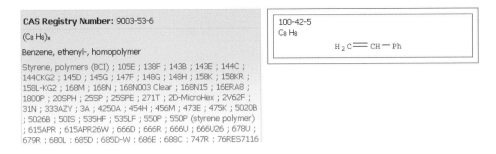

**CAS Registry Number:** 9003-53-6

(C₈ H₈)ₓ

Benzene, ethenyl-, homopolymer

Styrene, polymers (8CI) ; 105E ; 138F ; 143B ; 143E ; 144C ; 144CKG2 ; 145D ; 145G ; 147F ; 148G ; 148H ; 158K ; 158KR ; 158L-KG2 ; 168M ; 168N ; 168N003 Clear ; 168N15 ; 16ERA8 ; 1800P ; 20SPH ; 25SP ; 25SPE ; 271T ; 2D-MicroHex ; 2V62F ; 31N ; 333AZY ; 3A ; 4250A ; 454H ; 456M ; 473E ; 475K ; 5020B ; 5026B ; 50IS ; 535HF ; 535LF ; 550P ; 550P (styrene polymer) ; 615APR ; 615APR26W ; 666D ; 666R ; 666U ; 666U26 ; 678U ; 679R ; 680L ; 685D ; 685D-W ; 686E ; 688C ; 747R ; 76RES7116

100-42-5

C₈ H₈

$H_2C = CH - Ph$

***Figure A4.27***

Polymers with different tacticity (stereochemistries) are indexed separately; e.g. CAS Registry Numbers 9003-07-0, 25085-53-4, and 26063-22-9 apply to the atactic, isotactic, and syndiotactic polypropylenes respectively. Post-treated polymers, block polymers, graft polymers, and polymer blends are indexed in different ways and an example is given in Figure 6.14.

### A4.4.1    Homopolymers

The molecular formula for a homopolymer contains the formula for the monomer in parentheses followed by suffix 'x'. An example is shown in Figure A4.27.

*A4.4.1.1    Notes*

1. Only some of the names for polystyrene are shown.
2. Homopolymers may be searched easily through molecular formulas or through Exact search, in which the Search Option: Polymers is checked.

### A4.4.2    Copolymers

The molecular formula for a copolymer contains the formula for the monomers in parentheses followed by suffix 'x'. An example is shown in Figure A4.28.

*A4.4.2.1    Notes*

1. Only some of the names for ABS are shown.
2. Copolymers may be searched easily through structure queries with the separate components or through searches based on formulas.

### A4.4.3    Structure Repeating Units

An example of registration of a polymer as a structure repeating unit is shown in Figure A4.29.

*A4.4.3.1    Note*

1. See Section 6.10 for a further discussion on structure repeating units.

CAS Registry Number: 9003-56-9

(C₈ H₈ . C₄ H₆ . C₃ H₃ N)ₓ

2-Propenenitrile, polymer with 1,3-butadiene and ethenylbenzene

Acrylonitrile, polymer with 1,3-butadiene and styrene (8CI) ;
Acrylonitrile, polymer with butadiene and styrene (6CI) ; 1,3-
Butadiene polymer, with acrylonitrile and styrene (6CI) ; 1,3-
Butadiene, polymer with ethenylbenzene and 2-propenenitrile
(9CI) ; Benzene, ethenyl-, polymer with 1,3-butadiene and 2-
propenenitrile (9CI) ; Styrene, polymer with acrylonitrile and 1,3-
butadiene (8CI) ; Styrene, polymer with acrylonitrile and
butadiene (6CI) ; 0215A ; 06-10A ; 10JK2 ; 15NP ; 2020AST ;
2501K ; 3001M ; 300SF ; 301K ; 342EZ ; 429J ; 480S ; 660SF ;
747S ; 750A ; 750SW ; 757K ; 759A ; 88K4 ; 9715A ; 9738R ;
975BK1 ; 9815A ; A 201 ; A 201 (styrene polymer) ; A 402 ; A
404 ; A 404 (polymer) ; A 4571S27 ; A 50B ; ABS ; ABS
(polymer) ; ABS 1 ; ABS 10 ; ABS 12 ; ABS 130 ; ABS 150 ; ABS
170 ; ABS 180 ; ABS 200NT ; ABS 2020 ; ABS 2501K ; ABS 350 ;
ABS 380 ; ABS 4 ; ABS 400 ; ABS 433 ; ABS 547P ; ABS 55 ;
ABS 55NP ; ABS 60 ; ABS 606 ; ABS 680 ; ABS 707 ; ABS 750SW
; ABS 757 ; ABS 900 ; ABS 9815 ; ABS N-WN ; ABS copolymer ;
ABS plastic ; ABS resin ; ABS terpolymer ; ABS-BKWB ; ABS-C 08

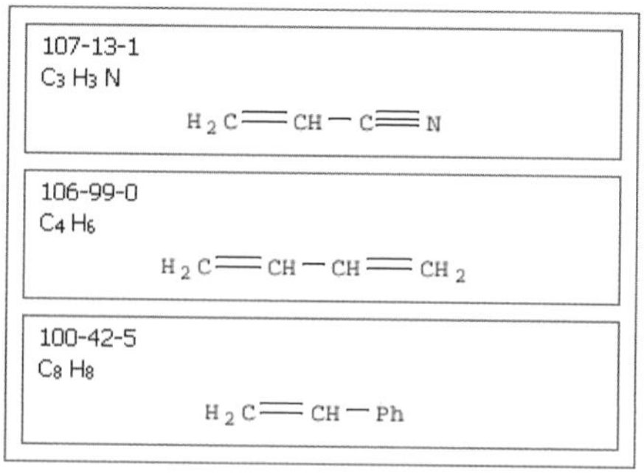

---

107-13-1
C₃ H₃ N

$$H_2C = CH - C \equiv N$$

---

106-99-0
C₄ H₆

$$H_2C = CH - CH = CH_2$$

---

100-42-5
C₈ H₈

$$H_2C = CH - Ph$$

*Figure A4.28*

---

CAS Registry Number: 28087-45-8

(C₁₂ H₁₂ O₄)ₙ

Poly(oxy-1,4-butanediyloxycarbonyl-1,3-phenylenecarbonyl)

Poly(oxytetramethyleneoxyisophthaloyl) (8CI) ; 1,4-Butanediol-
dimethyl isophthalate polymer, SRU ; 1,4-Butanediol-divinyl
isophthalate copolymer, SRU ; 1,4-Butanediol-isophthalic acid
copolymer, sru ; 1,4-Butanediol-isophthaloyl chloride copolymer,
SRU ; 1,4-Butanediol-isophthaloyl chloride polymer, sru ; 1,4-
Butylene glycol-isophthalic acid copolymer, SRU ; 1,4-
Dibromobutane-dipotassium isophthalate copolymer, SRU ; 1,4-
butanediol-dimethyl isophthalate copolymer, SRU ; Poly(butylene
isophthalate), SRU ; Poly(tetramethylene isophthalate), sru

Polymer

**Polymer Class Term:** Polyester

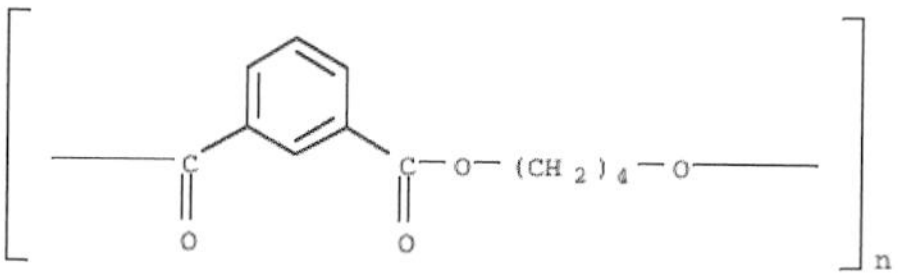

*Figure A4.29*

### A4.4.4   Proteins

CAS register all substances as precisely as possible.  Accordingly, if two protein sequences differ by even one amino acid, then different CAS Registry Numbers are used (see Figures A4.30 and A4.31). See Section 6.9 for a further discussion on registration of proteins. Additional information is available at http://www.cas.org/expertise/cascontent/registry/sequences.html.

### A4.4.5   Nucleic Acids and Related Substances

If two nucleic acid sequences differ by even one nucleic acid base, then different CAS Registry Numbers are used (see Figures A4.32 and A4.33). This has considerable implication in the rapidly developing area of molecular biology, and searchers should be particularly careful to ensure that all appropriate CAS Registry Numbers are retrieved.

**CAS Registry Number:** 33507-63-0

C63 H98 N18 O13 S

Substance P

1: PN: US20020037833 SEQID: 1 unclaimed sequence ; 21: PN: WO0181408 SEQID: 44 claimed protein ; 2: PN: JP2005049164 SEQID: 2 claimed protein ; 36: PN: WO2007058336 SEQID: 36 claimed protein ; 44: PN: WO2005016244 PAGE: 68 claimed protein ; 690: PN: WO20C4005342 PAGE: 46 claimed protein ; L-Methioninamide, L-arginyl-L-prolyl-L-lysyl-L-prolyl-L-glutaminyl-L-glutaminyl-L-phenylalanyl-L-phenylalanylglycyl-L-leucyl- ; Neurokinin P ; Substance P (1-11) ; Substance P (peptide) ; Substance P (smooth-muscle stimulant)

Component

**Protein Sequence**
Sequence Length:  11
modified

**Deleted CAS Registry Numbers:** 11035-08-8; 12769-48-1

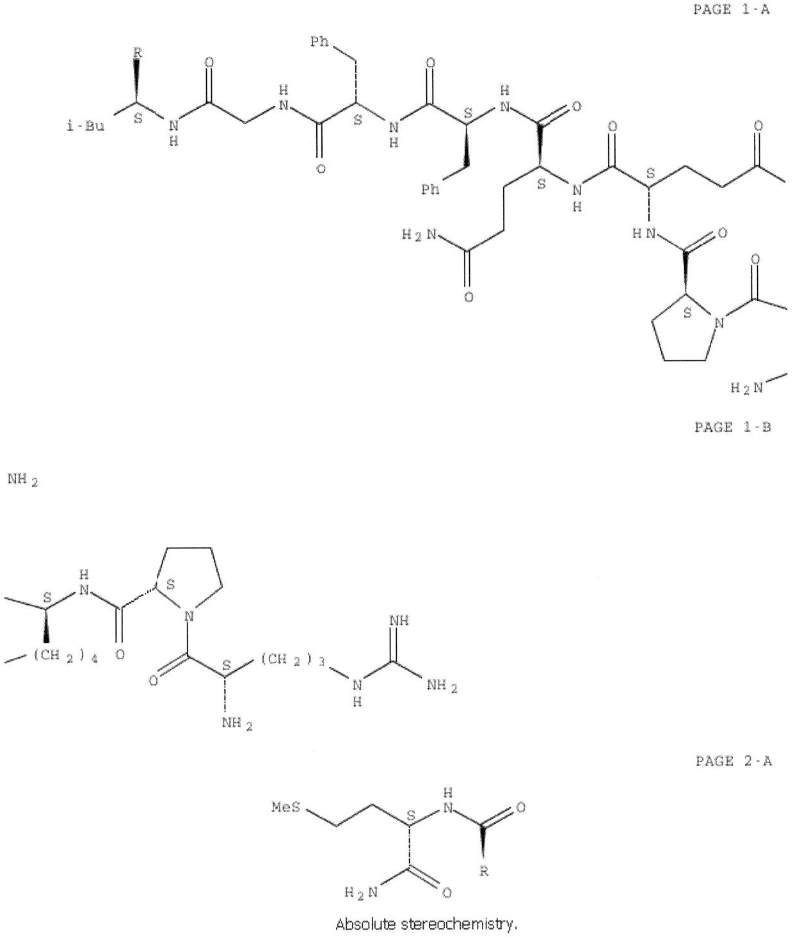

PAGE 1-A

PAGE 1-B

PAGE 2-A

Absolute stereochemistry.

*Figure A4.30*

**CAS Registry Number:** 9015-68-3

Unspecified

Asparaginase

Colaspase ; Crasnitin ; Crisantaspase ; E.C. 3.5.1.1 ; Elspar ;
Erwinase ; Kidrolase ; L-Asnase ; L-Asparaginase ; L-Asparagine
amidohydrolase ; Leunase ; MK 965 ; NSC 109229 ; Oncospar ;
α-Asparaginase

Manual Registration

**Deleted CAS Registry Numbers:** 9037-33-6; 9037-34-7;
9060-77-9

No Structure Diagram Available

*Figure A4.31*

**CAS Registry Number:** 147178-97-0

Guanosine, 2'-deoxycytidylyl-(3'→5')-2'-deoxyguanylyl-(3'→5')-2'-
deoxycytidylyl-(3'→5')-2'-deoxyguanylyl-(3'→5')-2'-
deoxyadenylyl-(3'→5')-2'-deoxyadenylyl-(3'→5')-thymidylyl-
(3'→5')-thymidylyl-(3'→5')-2'-deoxycytidylyl-(3'→5')-2'-
deoxyguanylyl-(3'→5')-2'-deoxycytidylyl-(3'→5')-2'-deoxy-,
double-stranded complementary (9CI)
Component

**Nucleic Acid Sequence**
Sequence Length:  12
2 a 4 c 4 g 2 t

**Deleted CAS Registry Numbers:** 159814-54-7

Source of Registration:  CA

**Document Types:** Conference,  Dissertation,  Journal,  Patent

**Sequence:**

1   cgcgaattcg cg

*Figure A4.32*

## A4.5    Other Cases

### A4.5.1    Incompletely Defined Substances

Incompletely defined substances are those that have a known molecular formula but for
which the complete valence bond structure was not fully described in the original article.
For example, while $o-$, $m-$, and $p$-xylene are the specific dimethylbenzenes, if only
'xylene' is mentioned in the original article, then the incompletely defined substance
(Figure A4.34) is indexed. Similar issues are encountered with salts in which ions have
different possible ratios (Figure A4.35).

**CAS Registry Number:** 356111-17-6

Transcription factor Tfam (Rattus norvegicus strain Sprague-Dawley clone 1 precursor) (9CI)

GenBank BAA77755 ; GenBank BAA77755 (Translated from: GenBank AB014089)

**Protein Sequence**
Sequence Length: 244

**Genbank (R) Definitions and Features:**

**Accession Number:** AB014089

**Version Number:** AB014089.1 GI:4877352

**Definition:** Rattus norvegicus mRNA for mitochondrial transcription factor A (r-mtTFA), complete cds.

**Organism:** Rattus norvegicus;  Eukaryota; Metazoa; Chordata; Craniata; Vertebrata; Euteleostomi; Mammalia; Eutheria; Euarchontoglires; Glires; Rodentia; Sciurognathi; Muroidea; Muridae; Murinae; Rattus

Source of Registration:  CA

**Document Type:** Journal

**Feature Table**

| Feature Key | Location | Qualifier |
|---|---|---|
| source | 1..1463 | /organism="Rattus norvegicus"<br>/mol-type="mRNA"<br>/strain="Sprague-Dawley"<br>/db-xref="taxon:10116"<br>/clone="clones 1, 4, 6, 12, and 15."<br>/tissue-type="Brain"<br>/clone-lib="cDNA cloned into the Lambda ZAP vector (STRATAGENE)"<br>/dev-stage="Adult" |
| CDS | 106..840 | /codon-start=1<br>/product="mitochondrial transcription factor A (r-mtTFA)"<br>/protein-id="BAA77755.1"<br>/db-xref="GI:4877353"<br>/translation="MALFRGMWGVLRTLGRTGVEMCAGCGGRIPSPVSLICIPKCFSS LGNYPKKPMSSYLRFSTEQLPKFKAKHPDAKVSELIRKIAAMWRELPEAEKKVYEADF KAEWKVYKEAVSKYKEQLTPSQLMGLEKEARQKRLKKKAQIKRRELILLGKPKRPRSA YNIYVSESFQGAKDESPQGKLKLVNQAWKNLSHDEKQAYIQLAKDDRIRYDNEMKSWE EQMAEVGRSDLIRRSVKRPPGDISEN" |
| sig-peptide | 106..231 | |
| polyA-site | 1463 | /note="15 a nucleotides" |

**Sequence:**

```
  1  MALFRGMWGV LRTLGRTGVE MCAGCGGRIP SPVSLICIPK CFSSLGNYPK
 51  KPMSSYLRFS TEQLPKFKAK HPDAKVSELI RKIAAMWREL PEAEKKVYEA
101  DFKAEWKVYK EAVSKYKEQL TPSQLMGLEK EARQKRLKKK AQIKRRELIL
151  LGKPKRPRSA YNIYVSESFQ GAKDESPQGK LKLVNQAWKN LSHDEKQAYI
201  QLAKDDRIRY DNEMKSWEEQ MAEVGRSDLI RRSVKRPPGD ISEN
```

*Figure A4.33*

**CAS Registry Number:** 1330-20-7

C8 H10

Benzene, dimethyl-

Xylene (8CI) ; Dilan ; Dimethylbenzene ; Xylol ; ZEP-RD

Incompletely Defined Substance Component

**Deleted CAS Registry Numbers:** 8026-09-3

2   ( D1 — Me  )

*Figure A4.34*

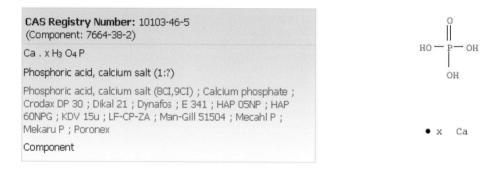

**CAS Registry Number:** 10103-46-5
(Component: 7664-38-2)

Ca . x H3 O4 P

Phosphoric acid, calcium salt (1:?)

Phosphoric acid, calcium salt (8CI,9CI) ; Calcium phosphate ;
Crodax DP 30 ; Dikal 21 ; Dynafos ; E 341 ; HAP 05NP ; HAP
60NPG ; KDV 15u ; LF-CP-ZA ; Man-Gill 51504 ; Mecahl P ;
Mekaru P ; Poronex

Component

*Figure A4.35*

*A4.5.1.1   Note*

1. Generally, incompletely defined substances will be retrieved in substructure searches when the structure searched is part of the incomplete structure for the substance in the database.

## A4.5.2   Minerals

Naturally occurring minerals have many different chemical compositions or crystalline forms and may be registered in a variety of ways (see Figures A4.36 and A4.37).

*A4.5.2.1   Note*

1. Many minerals are easily retrieved through name searches, either directly in REGISTRY or through **Explore References:  Research Topic**.

## A4.5.3   Records with 'No References'

Records with 'no references' may appear because of a number of reasons, including:

• manual registration of substances, which may occur because companies need CAS Registry Numbers for commercial purposes;
• CAS registrations for parent ring systems (Figure A4.38);

CAS Registry Number: 14940-68-2
(Component: 10193-36-9)

H₄ O₄ Si . Zr

Zircon (Zr(SiO4))

Zircon (8CI) ; A-DAX ; A-PAX 45M ; Standard SF 200 ; Ultrox
500W ; Zircobit MO ; Zircon Flour ; Zircon Flour 350 ; Zirconite ;
Zircosil 15 ; Zircosil 5

Mineral Component

● Zr(IV)

*Figure A4.36*

CAS Registry Number: 1318-74-7

Al . H O . O₅ Si₂
Al₂ H₄ O₉ Si₂

Kaolinite (Al2(OH)4(Si2O5))

Kaolinite (7CI,8CI) ; Kaolinite (Al2(Si2O7).2H2O) ; 50R ; 50R
(mineral) ; ASP 072 ; Argirec B 24 ; Asiacoat HG ; BIOFIX SC ;
Barnett Clay ; Biofix C 1 ; Biofix C 2 ; Biofix E 1 ; Biofix E 2 ;
Continental Clay ; GK ; GK (mineral) ; HG 90 ; Hard Top Clay S ;
Huber Clay HG 90 ; Hydrite 121 ; Hydrite MP ; Hydrite PX ;
Hydrite PXN ; Hydrite R ; Hydrite flat D ; Kaopaque 10 ;
Kaopaque 10S ; Kaopaque 20 ; Kaopaque 30 ; Kaophile 2 ;
Kaophobe 45 ; Kingwhite 65 ; MC 6J ; Mono 90 ; PP 0559 ; Polyfil
HG 90 ; SPS Kaolin ; UG 90

Mineral Component Tabular Inorganic Substance

**Deleted CAS Registry Numbers:** 11120-39-1; 12377-03-6;
39317-10-7; 51700-52-8; 51700-53-9; 51700-54-0; 51700-55-1;
52108-97-1; 60194-16-3; 61026-82-2; 62163-80-8; 66649-37-4;
90803-82-0; 188570-76-5; 239478-72-9

*Figure A4.37*

CAS Registry Number: 21426-57-3

C₈ H₁₀

Tetracyclo[3.3.0.02,4.06,8]octane (8CI,9CI)
Ring Parent

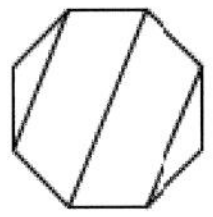

*Figure A4.38*

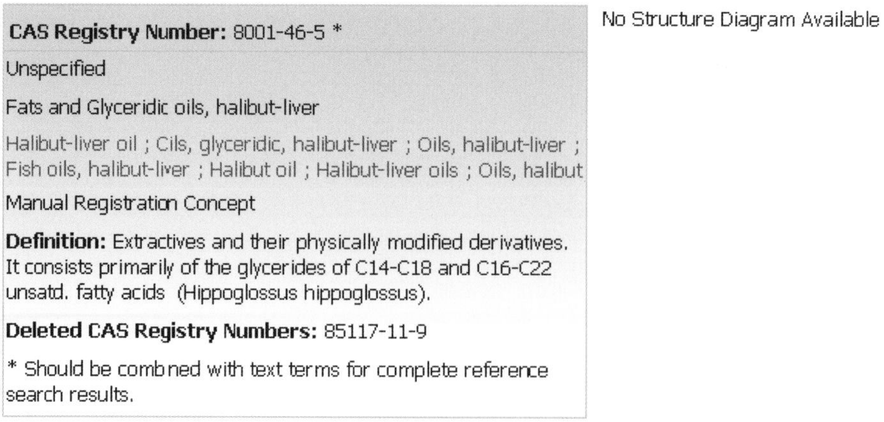

**CAS Registry Number:** 8001-46-5 *

Unspecified

Fats and Glyceridic oils, halibut-liver

Halibut-liver oil ; Oils, glyceridic, halibut-liver ; Oils, halibut-liver ;
Fish oils, halibut-liver ; Halibut oil ; Halibut-liver oils ; Oils, halibut

Manual Registration Concept

**Definition:** Extractives and their physically modified derivatives.
It consists primarily of the glycerides of C14-C18 and C16-C22
unsatd. fatty acids  (Hippoglossus hippoglossus).

**Deleted CAS Registry Numbers:** 85117-11-9

* Should be combined with text terms for complete reference
search results.

No Structure Diagram Available

*Figure A4.39*

- because the substance database may have been updated before the bibliographic database (and hence the CAS Registry Number has been assigned but the complete bibliographic record has not yet appeared);
- CAS does not use the CAS Registry Number for indexing in the bibliographic database, which commonly occurs for crude natural product extracts (Figure A4.39). 'Substances' of the latter type should be searched under **Explore References: Research Topic** using the Index Name.

# Appendix 5

# Understanding Structure Searches

Structures in REGISTRY have valence bond representations, and hence have single, double, and triple bonds. However, bonds in 'resonance' and 'tautomeric' situations (as defined in the following text) are specified as 'normalized bonds', which effectively means '*either* single or double bonds'.

## A5.1   The Resonance Issue

Bonds in substances in which valence bond structures have alternating single and double bonds in rings with *even numbers of atoms* are defined as 'normalized'. Note that the definition does not exactly relate to 'aromatic' compounds since cyclobutadiene (an antiaromatic compound) is defined in REGISTRY with normalized bonds (Figure A5.1), whereas thiophene (an aromatic compound) is defined with single and double bonds (since there are an odd number of atoms in the ring) (Figure A5.2).

## A5.2   The Tautomerism Issue

Where the structure representation in Figure A5.3 occurs, the bonds are defined as 'normalized'. Effectively, X, Y, and Z may be almost any atom, although all three cannot be carbon (simple alkenes are registered with double bonds!). The definition has wide implications and is most commonly encountered with keto/enols and with carboxylic acid derivatives. A key aspect is the hydrogen atom, so, for example, carboxylic acids and primary and secondary amides have 'normalized' bonds, although carboxylic acid esters and tertiary amides have exact single and double bonds.

The implications for the searcher are not important at the structure input stage, but often the searcher needs to understand the issue in order to interpret why certain answers have been retrieved. **Show precision analysis** may be needed to narrow the answers.

*Information Retrieval: SciFinder®, Second Edition* Damon D. Ridley
© 2009 John Wiley & Sons, Ltd

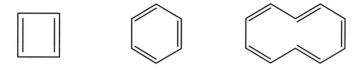

**Figure A5.1**   Examples of structures with normalized bonds

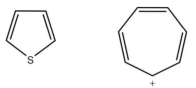

**Figure A5.2**   Examples of structures with exact bonds (single and double)

**Figure A5.3**   Normalized bonds in tautomers

## A5.3   Chain Lock Tool

The structure connection tables in REGISTRY specifically label bonds as either chain or ring bonds. In substructure searches, SciFinder applies by default a chain *or* ring value to any chain bond drawn in the query. The chain-locking tool overrides the default and ensures that the chain bond drawn only retrieves chain bonds in answers.

## A5.4   Ring Lock Tool

The structure inputs in REGISTRY specifically give all ring atoms an additional label 'D' or 'T'(Figure A5.4). The 'D' is used when the atom is part of a single ring (i.e. has two ring bonds only), whereas 'T' is used when the atom has three or more ring bonds.
    In structure queries, by default, SciFinder allows either 'D' or 'T' values to ring atoms drawn with two ring bonds (i.e. the left-hand structure *query* would have two atoms 'T' and eight atoms 'D' or 'T'). However, when the Ring Lock tool is applied, ring atoms with two ring bonds are specified 'D' values only (i.e. the left-hand structure *query* would have eight atoms specified 'D' and two atoms specified 'T').

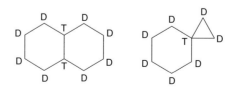

**Figure A5.4**   Ring atom descriptions for actual structures in REGISTRY

# Appendix 6

# Original Publication Discussed in Chapter 7, Section 7.1

1676                          *J. Am. Chem. Soc.* **1997**, *119*, 1676−1681

## Ester Hydrolysis by a Catalytic Cyclodextrin Dimer Enzyme Mimic with a Metallobipyridyl Linking Group

**Biliang Zhang and Ronald Breslow\***

*Contribution from the Department of Chemistry, Columbia University, New York, New York 10027*

*Received October 29, 1996*[®]

**Abstract:** A $\beta$-cyclodextrin dimer with a linking bipyridyl group is synthesized as a catalyst precursor, a holoenzyme mimic. It binds both ends of potential substrates into the two different cyclodextrin cavities, holding the substrate ester carbonyl group directly above a metal ion bound to the bipyridyl unit. The result is very effective ester hydrolysis with good turnover catalysis. For example, a Cu(II) complex accelerates the rate of hydrolysis of several nitrophenyl esters by a factor of $10^4$−$10^5$, with at least 50 turnovers and no sign of product inhibition. In the best case, with an added nucleophile that also binds to the metal ion, a rate acceleration of $1.45 \times 10^7$ over the background reaction rate was observed. Hydrolysis by a catalyst with only one cyclodextrin binding group is significantly slower than in the bidentate binding cases. As expected, the binding of a transition state analogue to these catalysts is stronger with the metal ion present than without. This and kinetic evidence point to a mechanism in which the metal ion plays a bifunctional acid−base role, enforced by the binding geometry that holds the substrate functionality right on top of the catalytic metal ion.

**Scheme 1**

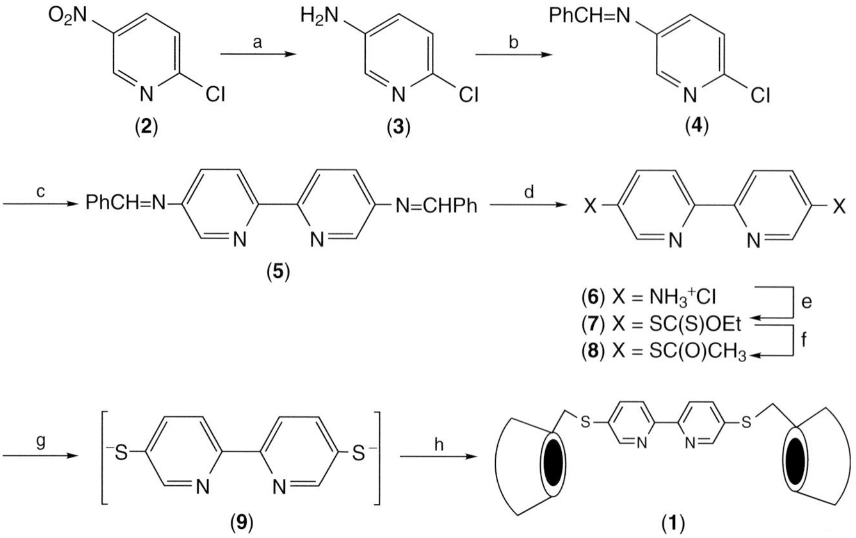

(2)    (3)    (4)

(5)

(6) X = NH₃⁺Cl
(7) X = SC(S)OEt
(8) X = SC(O)CH₃

(9)    (1)

(a) Fe/NH₄Cl, MeOH, H₂O, rt: (b) PhCHO, MgSO₄, Et₃N, CH₂Cl₂, rt. 24 h: (c) NiBr₂(PPh₃)₂, Zn/Et₄N⁺I⁻/THF, 50–80 °C, 20 h: (d) I N HCl, reflux; (e) (1) NaNO₂/H₃⁺O, (2) KSC(S)OEt, H₂O, 65–70 °C; (f) (1) 20% KOH/EtOH, reflux, (2) CH₃COCl, 0–5 °C; (g) NH₃/MeOH, rt I h: (h) mono-6-iodo-beta-cyclodextrin, DMF, 60–65 °C. 3 h

**Scheme 2**

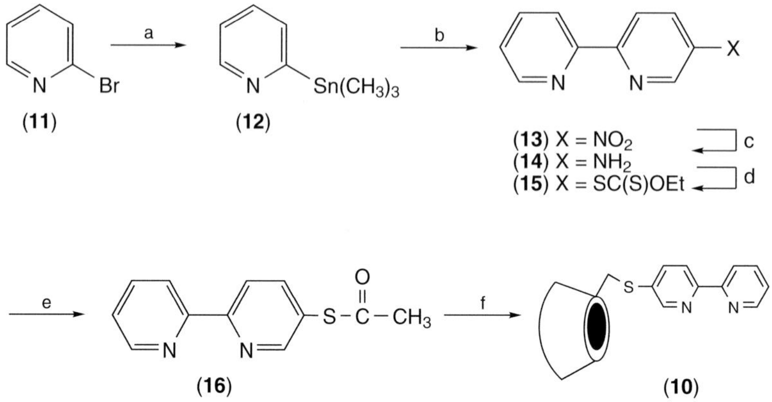

(11)    (12)

(13) X = NO₂
(14) X = NH₂
(15) X = SC(S)OEt

(16)    (10)

(a) (1) BuLi/EtO, −78 °C, (2) (CH₃)₃SnCl, THF, −78 °C; (b) 2-chloro-5-nitropyridine, Pd(Ph₂P)₂Cl₂, THF, reflux, 24 h: (c) pd° on activated carbon(10%), NaBH₄, MeOH, rt 5 h: (d) (I) NaNO₂/H₃⁺O, (2) KSC(C)OEt, H₂O, 65–70 °C; (e) (1) 20% KOH/EtOH, reflux; (2) CH₃COCl, 0–5 °C; (f) (1) NH₃/MeOH, rt, I h; (2) mono-6-iodo-beta-cyclodextrin, DMF, 60–65 °C,3 h

# Index

*Information Retrieval: SciFinder®, Second Edition* Damon D. Ridley
© 2009 John Wiley & Sons, Ltd